2nd International Conference on Advances in Materials, Mechanical and Manufacturing (AMMM 2020)

IOP Conference Series: Materials Science and Engineering
Volume 893

Online
27 - 29 March 2020

ISBN: 978-1-7138-2690-3
ISSN: 1757-8981

Printed from e-media with permission by:

Curran Associates, Inc.
57 Morehouse Lane
Red Hook, NY 12571

Some format issues inherent in the e-media version may also appear in this print version.

Institute of Physics
Dirac House, Temple Back
Bristol BS1 6BE UK

Phone: 44 1 17 929 7481
Fax: 44 1 17 920 0979

techtracking@iop.org

Additional copies of this publication are available from:

Curran Associates, Inc.
57 Morehouse Lane
Red Hook, NY 12571 USA
Phone: 845-758-0400
Fax: 845-758-2633
Email: curran@proceedings.com
Web: www.proceedings.com

TABLE OF CONTENTS

MATERIAL SYNTHESIS AND PREPARATION

MECHANICAL PROPERTIES AND AUTOMATION EQUIPMENT

IOP Conf. Series: Materials Science and Engineering **893** (2020) 011001 doi:10.1088/1757-899X/893/1/011001

Preface

We are very pleased to introduce the proceedings of the 2020 the 2nd International Conference on Advances in Materials, Mechanical and Manufacturing (AMMM 2020), which was held in full virtual style via ZOOM during March 27-29, 2020.

Taking into account worldwide travel restrictions due to the COVID-19 pandemic, the organizing committees decided, after careful consideration and advices from participants, to hold AMMM 2020 as a fully virtual conference. Although this is a very difficult decision for us, it also gives an opportunity to innovate and test a new style of conference by a yet safer way in this specific period. The safety and well-being of participants is of paramount importance for us all the time.

The Conference has been an opportunity for participants coming from different countries to present and discuss their topics in their respective research areas. We would like to thank all participants for their contributions to the conference program and for their contributions to this proceedings. Our special thanks go to our Conference Committees and reviewers for their devoted assistance in the overall organization of the conference. We also appreciate all speakers and participants' understanding about the conference style change and support to AMMM conference all the time.

We are looking forward to the next International Conference on Advances in Materials, Mechanical and Manufacturing. We hope to meet everyone in AMMM 2021 conference at the site again

AMMM Organizing Committees

IOP Conf. Series: Materials Science and Engineering **893** (2020) 011002 doi:10.1088/1757-899X/893/1/011002

Conference Committees

International Advisory Committees

Andrew Y C Nee, National University of Singapore, Singapore

Mingwang Fu, The Hong Kong Polytechnic University, Hong Kong

Conference Chair

Sergei Alexandrov, Beihang University, China & Institute for Problems in Mechanics of the Russian Academy of Science, Russia

Program Chairs

Zengtao Chen, University of Alberta, Canada

Yeong-Maw Hwang, National Sun Yat Sen University, Taiwan

Ken-ichi Manabe, Tokyo Metropolitan University, Japan

Steering Committees

Renfu Li, Huazhong University of Science and Technology, China

Lang Lihui, Beihang University, China

Huafeng Ding, China University of Geosciences, China

Technical Committees

Joerg Niemann, University of Applied Sciences Dusseldorf, Germany

Aleksander Muc, Cracow University of Technology, Poland

Irena Markovska, Assen Zlatarov University, Bulgaria

Wei Cui, Northeast Petroleum University, China & Nanyang Technological University, Singapore

Chun IL Kim, University of Alberta, Canada

Andrew Rae, University of the Highlands and Islands, UK

João Manuel R. S. Tavares, Universidade do Porto (FEUP), Portugal

Hamza Blala, The Algerian Space Agency, Algeria & Beihang University, China

Jinqiang Ning, Georgia Institute of Technology, USA

Mohamed Arezki Mellal, M'Hamed Bougara University (UMBB), Algeria

En-Chih Chang, I-Shou University, Taiwan

Hasan AL Dabbas, Philadelphia University, Jordan

IOP Conf. Series: Materials Science and Engineering **893** (2020) 011002 doi:10.1088/1757-899X/893/1/011002

Peer review statement

All papers published in this volume of *IOP Conference Series: Materials Science and Engineering* have been peer reviewed through processes administered by the proceedings Editors. Reviews were conducted by expert referees to the professional and scientific standards expected of a proceedings journal published by IOP Publishing.

IOP Conf. Series: Materials Science and Engineering **893** (2020) 012001 doi:10.1088/1757-899X/893/1/012001

The Analysis about Synthesis, Structure and Properties of Willemite Ceramic Pigments Obtained by a Sol–Gel Method

Ts I Dimitrov[1], I G Markovska[2] and Ts H Ibreva[2]

[1] Russe University Angel Kanchev, Razgrad Branch, Bulgaria
[2] Prof. Assen Zlatarov University, Burgas, Bulgaria

Abstract. This work focuses on the syntheses of willemite ceramic pigments doped with cobalt and nickel ions (Co^{2+}, Ni^{2+}). The pigments were prepared by the sol–gel process. The phase formation has been studied by X - ray diffraction analysis. The results showed that phase composition depends on sintering temperature, isothermal retention period as well as type of chromophore ion. The basic phase is obtained at even 900 - 1000^{0}C. Color characteristics of the pigments have been measured using the Lovibond Tintometer RT 100 Color – CIELab color measurement. Better results were obtained at pigments synthesized at 1100^{0}C. High temperature ceramic pigments representing solid solution with stable crystalline phase of Zn_2SiO_4 were obtained. The best pigments have been added to white faience glaze and sanitary ceramics.

1. Introduction

Willemite, Zn_2SiO_4 with phenakite structure is an orthosilicate with all atoms in general position and composed by a framework of tetrahedra accommodating zinc and silicon in three different fourfold crystallographic sites: two slightly different zinc sites Zn1(1.950Å) and Zn2 (1.961Å), and Si (1.635 Å), in rhombohedral symmetry with lattice parameters a = b ~13.948Å, and c ~9.315 [1-5].

Many other researchers [6-9] have devoted efforts on topics related not only the chemistry of the process but also to finding the main fields of application of the sol-gel technology. The sol-gel method is related to hydrolysis and polycondensation of metal alcoholates $Me(OR)_n$, where Me-metal with valence of n and R-alkyl radical C_XH_{2X+1}. At temperatures close to that of the environment, hydrolysis of the solution of the metal alcoholates is carried out and then polycondensation takes place which gives sol. As the reaction this sol transforms into gel which is heated to obtain the end product.

2. Experiment

2.1. Raw materials

The synthesis of pigments is carried by sol-gel method using raw materials as TEOS - $Si(OC_2H_5)_4$, $Zn(NO_3)_2.6H_2O$, $Co(NO_3)_2.6H_2O$ and $Ni(NO_3)_2.6H_2O$ (pure grade).

For the preparation of willemite pigments by the sol-gel technology have been chosen the blends for which the best result were obtained with respect to color intensity after the solid state synthesis: $0.375CoO.1.625ZnO.SiO_2$ and $0.375NiO.1.625ZnO.SiO_2$. Due to the strong propensity of tetraethyl orthosilicate (TEOS) to hydrolysis, ethyl alcohol was added to it under continuous stirring. The nitrates of chlorides of the corresponding chromophore components were preliminarily dissolved in ethyl alcohol. After mixing the solutions, the mixture was intensely stirred with magnetic stirrer for 30 min for better homogenization.

IOP Conf. Series: Materials Science and Engineering **893** (2020) 012001 doi:10.1088/1757-899X/893/1/012001

2.2. Thermal treatment

The sol obtained transformed into gel-like mass after 48 h at 60°C and it was dried at temperature of 120°C, and then at 180°C, until complete separation of the nitrates. After the full drying, the samples were ground in agate mortar and subjected to thermal treatment at temperatures of 600°C, 700°C, 800°C, 900°C, 1000°C, 1100°C and 1200 °C with isothermal periods of 4 h.

2.3. Method of analysis

Phase composition of the synthesized ceramic pigments was determined using X-ray diffraction (XRD) with a Bruker D8 diffractometer operating at 40 kV and 40 mA with CuK_α radiation. FT-IR spectra were collected using a Tensor 37 spectrometer (Bruker) with a 4 cm^{-1} resolution on standard KBr pallets in the spectral region 400-4000 cm^{-1} at room temperature. The EPR spectra were taken with spectrophotometer Bruker EMX Premium X equipped with a system allowing varying the measurement temperature in the interval 120-450 K. The electron microscope photographs were taken using scanning electron microscope "JEOL 6390" with INCA Oxford analyst. The color determination of the pigments is determined spectrally by a tintometer of Lovibond Tintometer RT 100 Color.

3. Results and discussions

3.1. X-ray analyses

X-ray phase analysis is a direct method for phase identification. The method is based on X-ray diffraction. The main task of the XRD analysis is identification of the different phases separately or in mixtures on the basis of the diffraction pattern recorded with the sample studied.

The main method of phase analysis is with powdery sample which has been widely used due to its simplicity and universality. Diffraction patterns of the synthesized ceramic pigments are presented in figure 1 (A, B).

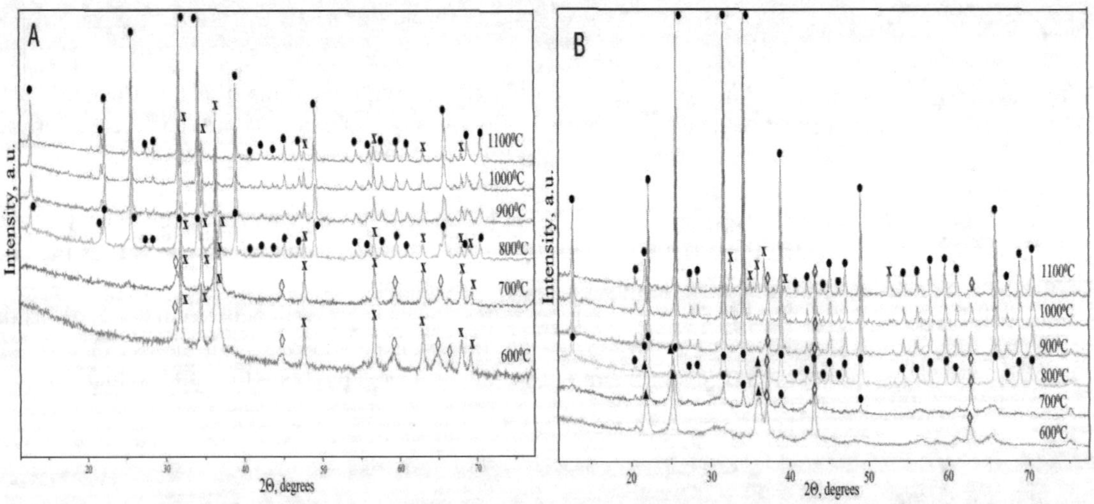

Figure 1. A) X – ray spectra of willemite pigments with composition 0,375CoO.1,625ZnO.SiO_2 synthesized at temperature: 600÷1000^{0}C, main phases: • – Co willemite ; x - zincite ZnO; ◊ - Co_3O_4
B) X-ray spectra of pigments with composition 0,375NiO.1,625ZnO.SiO_2 synthesized at temperature: 600 ÷1000 ^{0}C, with main phases: • – solid solution of Ni in willemite; x - Nickel Orthosilicate (Ni_2SiO_4); ◊ - bunsenit NiO; ▲-quartz SiO_2

Due to the size of ionic radii of Zn^{2+} - 0. 74 Å and Co^{2+} - 0.65 Å a solid solution is formed by incorporation of cobalt in the willemite lattice. Data for unit cell parameters presented by Ozel et al. [1] confirme that. It can be seen from the X-ray patterns shown in Figure 1 that the main phase Co-willemite was synthesized at temperature as low as 800°C and reflexes from cobalt oxide and zinc oxide were also registered. With the increase of the temperature of thermal treatment, a tendency of increase of the intensities of the peaks for the main phase Co-willemite accompanied by decrease of the peak intensities for ZnO. Similar results were obtained in the system 0,375NiO.1,625ZnO.SiO_2. The X- ray data well correlated with the data from the FT-IR spectroscopy.

3.2. FT- IR spectra of pigments

The FT- IR spectra presented in Figure 2 shown the presence of absorption maxima characteristic for the pure willemite (Table 1): the symmetric valent vibration at 868 (869) cm^{-1} is due to the presence of Si-O bands, 900, 931 and 977 cm^{-1} – asymmetric valent of Si-O and deformation vibration at 459 cm^{-1}. The bands at 576 cm^{-1} are due to symmetric valent /614 cm^{-1} – asymmetric valent / vibrations of Zn-O in ZnO_4.

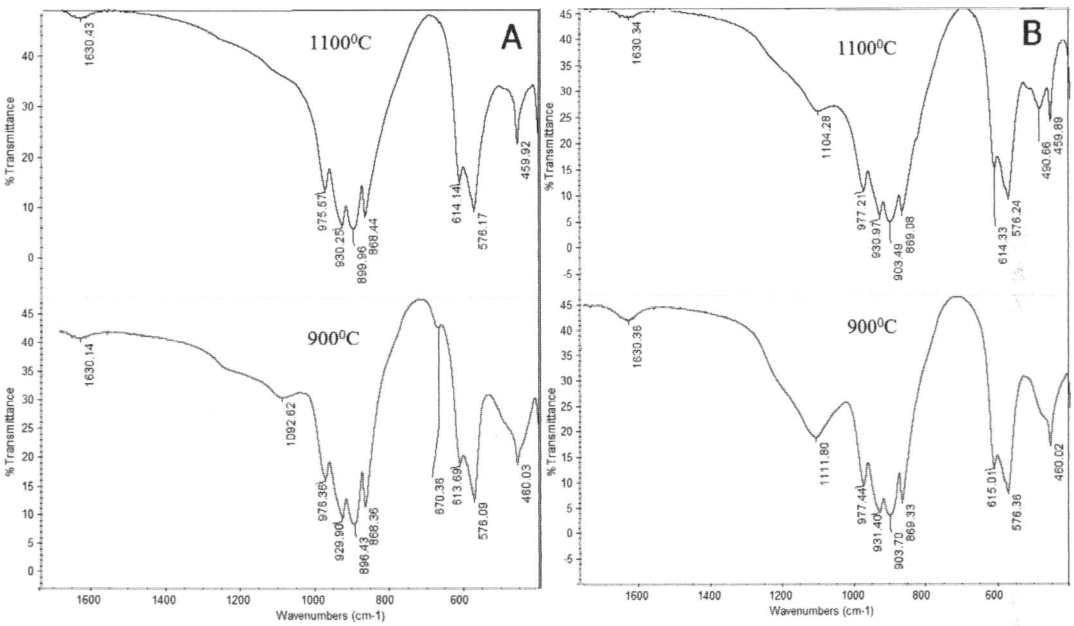

Figure 2. FT- IR spectra of pigments 0,375CoO.1,625ZnO.SiO_2(A) and 0,375NiO.1,625ZnO.SiO_2 (B)

Table 1. Basic IR - FT characteristic of willemite

Peak wave numbers (cm^{-1})	Type of oscillation, connection
977; 931; 900	(v_3) Si-O antisymetric stretching vibration
868, 869	(v_1) Si-O symmetric stretching
600-615	(v_3) Zn-O antisymetric stretching in ZnO_4
575	(v_1) Zn-O symmetric stretching in ZnO_4
459	(v_4) SiO_4 bending

3.3. Electron paramagnetic resonance

The EPR spectrum of pigment 0.375CoO.1.625ZnO.SiO_2 (1100°C) was registered in the temperature interval 100÷295 K (Figure 3).

IOP Conf. Series: Materials Science and Engineering **893** (2020) 012001 doi:10.1088/1757-899X/893/1/012001

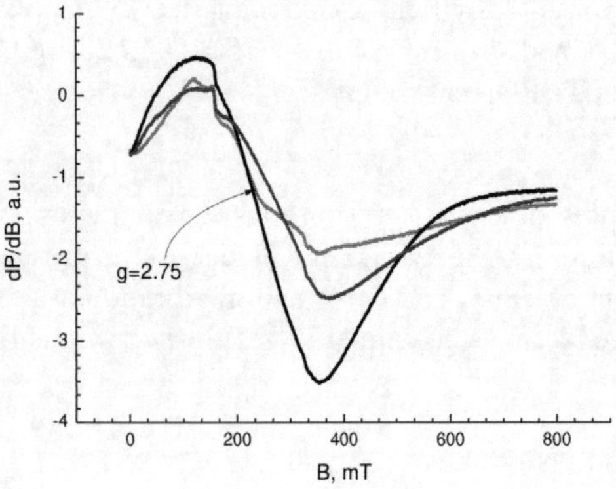

Figure 3. ESR spectrum of the sample $0.5CoO.1.5ZnO.SiO_2$ at 295 K (black curve), 210 K (blue curve) and 120 K (red curve).

The EPR spectra of sample $0.5CoO.1.5ZnO.SiO_2$ were registered in the temperature interval 120-295 K. A wide asymmetric signal with effective g-factor of 2.75 and width about 218 mT was observed at room temperature. With decrease of the temperature, this signal widens and at low temperatures, comparatively wider non-allowed overlapping signals were registered.

The single asymmetric signal observed at 295 K was attributed to electron exchange bonded Co^{2+} ions in field with octahedron distortion. The spectrum registered at 120 K contained lines of the Co^{2+} ions ultrafine structure (I=7/2) overlapping the wide signal which was attributed to exchange bonded Co^{2+} ions, as mentioned above. The ultrafine structure becomes visible at low temperatures due to the significant widening of the dominating signal at 295 K.

3.4. Color measuring

Color is one of the most important indicators of pigment quality. Colored substances absorb and convert light rays of a certain wavelength into the visible portion of the spectrum, due to their atomic structure. The CIELab system defines colors not only of ceramic pigments but also of other materials, which indicates that this system is universal and widely used. The colour measurements were performed using the CIELab method (Table 2). This method, which is the standard analysis in the ceramic industry, especially for the ceramic pigments allows to determine the whiteness and colour degree of tiles by measuring the three parameters: *L**, *a** and *b**, where: *L** (brightness), from absolute white *L** = 100 to absolute black *L**=0 ; a* - green color (-) / red color (+) , b* - blue color (-) / yellow color (+).

Table 2. Results of color coordinate measurements of pigments

№	Composition	Colour	L*	a *	b *
1	**$0,375CoO.1,625ZnO.SiO_2$ - 900°C**		**47,67**	**-10,96**	**-9,67**
2	**$0,375CoO.1,625ZnO.SiO_2$ - 1000°C**		**44,08**	**-9,57**	**-22,02**
3	**$0,375CoO.1,625ZnO.SiO_2$ - 1100°C**		**42,34**	**-3,82**	**-45,44**
4	**$0,375CoO.1,625ZnO.SiO_2$ - 1200°C**		**41,65**	**-0,65**	**-37,56**
5	**$0,375NiO.1,625ZnO.SiO_2$ - 900°C**		**74,83**	**-4,69**	**4,05**
6	**$0,375NiO.1,625ZnO.SiO_2$ - 1000°C**		**72,52**	**-8,58**	**-0,52**

7	**0,375NiO.1,625ZnO.SiO$_2$ - 1100°C**		**68,09**	**-14,40**	**-5,11**
8	**0,375NiO.1,625ZnO.SiO$_2$ - 1200°C**		**64,56**	**-12,17**	**-4,75**

It can be seen from the data presented that the color of the pigments synthesized with cobalt and zinc is blue and blue - green. With the increase of the sintering temperature, their luminance L* was found to decrease. The amount of blue color (- b*) was the highest for the pigment with composition 0,375CoO.1,625ZnO.SiO$_2$ synthesized at 1100°C, where (- b) = 45,44. This pigment (composition 3) is best suited for use in practice because it has the most intense blue color.

3.5. Scanning electron microscopy

The particle size of pigments can affect the final appearance of the coated surface. The morphology and microstructure of pigments were examined by scanning electron microscopy (SEM). As sputter coating material has been used gold, due to its high conductivity that enables high-resolution imaging. The ELMI Sputter installation was used for this purpose.

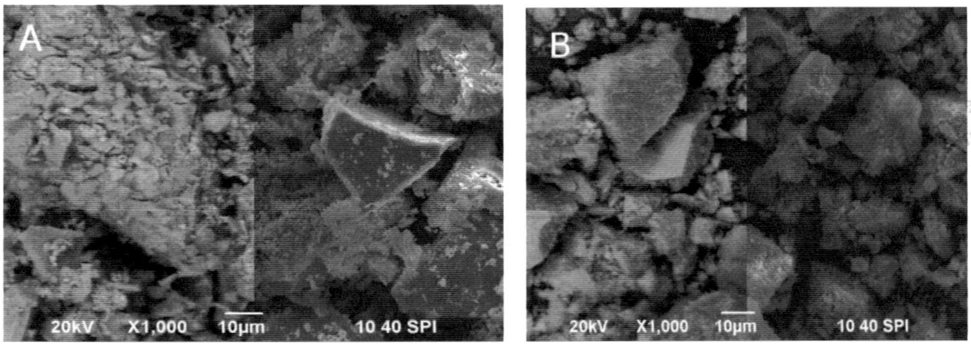

Figure 4. SEM of pigments with composition: A) 0,375CoO.1,625ZnO.SiO$_2$, B) 0,375NiO.1,625ZnO.SiO$_2$

The SEM micrographs of pigments containing Co^{2+}(composition 0,375CoO.1,625ZnO.SiO$_2$, 1100^{0}C) and Ni^{2+} (composition 0,375NiO.1,625ZnO.SiO$_2$, 1100^{0}C) are presented in Figure 4. Both pictures show that the pigments are monodispersed with unambiguous crystal habitus and crystal size of about 5-20µm.

4. Conclusions

Colored ceramic pigments were synthesized with participation of two chromophore ions: Co^{2+} and Ni^{2+}. The color of all pigments is blue and blue-green. Sintering process was carry out by sol-gel method. The optimal parameters of the synthesis process have been established. Due to the very similar ionic radii of Zn^{2+} - 0.60 Å and Co^{2+} - 0.58 Å a solid solution of substitution is formed. The best results were obtained with pigments synthesized at a firing temperature of 1100C. SEM micrographs show that the pigments are monodispersed with unambiguous crystal habitus and crystal size of about 5-20µm. The synthesized pigments are suitable and can be successfully applied in glazes for tiles and sanitary ceramics.

5. References

[1] Ozel E, Yurdakul H, Turan S, Ardit M, Cruciani G and Dondi M 2010 *Journal of the European Ceramic Society*, **30** 3319–3329
[2] Bragg W and Zachariasen W, *Z. Kristallogr.* 1930 **72**, 518–528
[3] Hang C, Simonov M and Belov N 1970 *Sov. Phys. Crystallogr.* **15** 387–390
[4] Klaska K, Eck J and Pohl D 1978 *Acta Cryst.* **B34** 3324–3325

[5] McMurdie H, Morris M, Evans E, B. Paretzkin B, Wong-Ng W, Hubbard C, *Powder Diffr.* 1986 **1**, 274
[6] Chen Z, Shi E, Li W, Zheng Y and Zhong W 2002 *Materials Letters* **55 (5)** 281-284.
[7] Chemlal S, Larbot A, Persin M, Sarrazin J, Sghyara M and Rafi M. 2000 *Materials Research Bulletin.* **14-15** 2515-2523
[8] Djambazov S, Ivanova Y and Yoleva A, 1998 *Ceram. Int.* **24** 281–284
[9] Cui H, Zayat M and Levy D 2005 *Journal of Sol-Gel Science and Technology* **35 (3)** 175-181

Acknowledgements
The financial support of this work by the Bulgarian Ministry of Education and Science, National Research Fund under the contract number KP-06-H27/14 - 2018 is gratefully acknowledged.

IOP Conf. Series: Materials Science and Engineering **893** (2020) 012002 doi:10.1088/1757-899X/893/1/012002

Influence of Physical Mixing Ratio on Pore Development in Highly Porous Carbon Prepared From Nipa Palm Husk Using Hydrothermal Carbonization with Chemical Activation

Sutthinan Dechabun[1], Parncheewa Udomsap[2], Nuwong Chollacoop[2] and Apiluck Eiad-ua[1]

[1] College of Nanotechnology, King Mongkut's Institute of Technology Ladkrabang, Bangkok, Thailand
[2] National Metal and Materials Technology Center, Pathumthani, Thailand

Corresponding author's email address: apiluck.ei@kmitl.ac.th

Abstract. In this study, porous carbon materials were prepared from nipa palm (*Nypa fruiticans*) husk using hydrothermal carbonization process with KOH activation through physical mixing. In order to investigate effects of factors on carbon products, hydrothermal temperature and duraton were set to 180°C and 8 hours, and KOH:hydrochar activation weight ratio at 1:1, 1:2 and 1:3, respectively. Prepared carbon materials then were carbonized at 900°C for 2 hours under nitrogen atmosphere and characterized by scanning electron microscope, fourier-transform infrared spectroscopy and x-ray diffraction. According to the results, carbon material produced using physical mixing activation ratio of 1:1 shows the most significant pore development and surface chemical properties, leads to finding a promising way to prepare environmental-friendly highly porous carbon material from biomass.

1. Introduction

Thailand is an agricultural country and has variety of plant species, leads to plenty of biomass available. Agricultural residues from plantation or orcharding are usually been thrown away or let them decompose over time, which yield little or no benefits from doing so. An alternative way to utilize large number of residues is enhancing their properties or changing their structures such as using as fuel[1,2] , direct fertilizers or as precursor for activated carbon[3].

Activated carbon has extraordinary porosity, mechanical and electrical properties[4]. Thus, it becomes an ideal material for adsorption[3,4,5,6], catalytic[7,8], structural and energy applications[1,2]. Activated carbon is mainly prepared from biomasses because they have organic structure with high carbon content called lignocellulosic components, consists of three vital types-cellulose, hemicellulose and lignin[9]. Pyrolysis used to be the main way to transform lignocellulosic components in biomass into char, yielding higher carbon content. However, pyrolysis method requires high energy to operate[10]. Hydrothermal carbonization process has gained in popularity as it requires less operation temperature than conventional pyrolysis. With hydrothermal carbonization method, biomass chemical decomposition can occur under controlled or self-generated pressure.[11] Water added to raw biomass helps in breaking chemical bonds in lignocellulosic components, removing excessive water and carboxyl group in biomass. Temperature and pressure promote reconstruction of organic compounds through polymerization, condensation and aromatization.[12]

Chemical activation is a method to prepare porous activated carbon. It has purposes to develop suitable porosity and surface chemical composition, leads to adsorption enhancement. Some substances such as sodium hydroxide (NaOH), potassium hydroxide (KOH) or phosphoric acid (H_3PO_4) are widely used as an activating agent.[13] Impregnation is a popular way as it uses activating agent solution to penetrate into biomass structure, yielding porosity development.[13] But according to P Nowicki's work on influences of chemical activation methods toward porosity of activated carbon prepared from pine cones shows that activated carbon prepared by physical mixing- another method of activation using solid activating agent- yields more porosity than one prepared by impregnation.[14] However, there are not many studies about suitable physical mixing ratios towards carbon material's structural development.

In this work, suitable physical mixing ratio was examined via mesoporous carbon powder preparation from nipa palm husk (*Nypa fruiticans*) using hydrothermal carbonization process along with chemical activation using KOH physical mixing. The aim on this work is to investigate the effect of physical mixing ratios of activating agent and hydrochar towards physical pore structure alternation. This research also has an object to introduce higher performance activation method to produce porous carbon materials with better porosity but use lower energy, reaction time and activation agent cost, leads to encouragement in more effective way to utilize biomasses to valuable products.

2. Materials and methods

2.1. Material
Nipa palm husks (*Nypa fruiticans*) are collected from local plantation source in Samut Songkhram, Thailand. Pelletized analysis-grade potassium hydroxide is obtained from Carlo Erba Reagents. Dry nipa palm husks were separated from shells, ground using blender and handheld grinder, and then sieved to obtain coarse powdered nipa palm husk (abbreviated as NPH).

2.2. Hydrothermal process
20g of NPH was loaded with 40ml of deionized water into Teflon vessel. Pack the closed vessel into stainless steel reactor and put in oven at 180°C for 8 hours under self-generated pressure. After finishing the hydrothermal process, the reactor was rapidly cooled down to halt the reaction using water. Hydrochar was removed from vessel and then was dried in oven at 90°C for 24 hours.

2.3. Activation using physical mixing
Pelletized KOH was ground and mixed with dried hydrochar with weight ratios between KOH and hydrochar as 1:1, 1:2 and 1:3, respectively. (4g of hydrochar was used. 4g, 2g and 1.33g of KOH were used.) Mixing was carried out for at least 15 minutes. Finally, the mixture was dried in oven at 90°C.

2.4. Carbonization
Dried activated mixture was loaded in a ceramic boat and inserted into horizontal tube furnace. Carbonization was done at 900°C for 2 hours, with heating rate of 10°C/min under continuous nitrogen gas atmosphere. (flow rate 100m^3/min) The resulting carbon material was washed with deionized water to neutralize pH and was dried in oven at 90°C for 90 minutes so it completely dried. Prepared samples were denoted as HTC180-08-X:Y while X:Y is each physical mixing ratio.

2.5. Characterization
Morphology of prepared porous carbon were investigated using scanning electron microscope (Zeiss EVO MA10). Gold sputtering was done to enable conductive surface operation. 10kV electron acceleration voltage was used and magnification scale is 1000x.

Fourier transform infrared spectroscopy technique (Perkin Elmer Spectrum Two) was used to observe surface chemical properties. Scanning wavenumber range is 4000-400cm^{-1} in ART mode. FT-IR spectra was reported as %T.

X-ray diffraction technique (SmartLab Rigaku) was used for investigating carbon formation in prepared carbon materials. $Cu_K\alpha$ was used as an x-ray source, theta-2-theta mode was used for every measurement and scanning angle range was 20-90°.

IOP Conf. Series: Materials Science and Engineering **893** (2020) 012002 doi:10.1088/1757-899X/893/1/012002

3. Results and discussion

3.1. Morphology of prepared porous carbon from nipa palm husks

(a) NPH

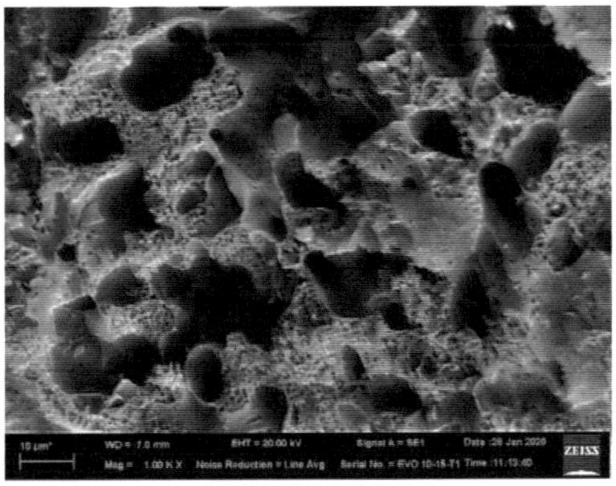

(b) HTC180-08-1:1

(c) HTC180-08-1:2

(d) HTC180-08-1:3

Figure 1. SEM images of (a) raw nipa palm husk powder, (b) HTC180-08-1:1, (c) HTC180-08-1:2 and (d) HTC180-08-1:3.

SEM images show the significant effect of activating agent which initiate alternations of physical structures in hydrochar prepared from nipa palm husk powder. When compares morphology image of raw nipa palm husk powder in Figure 1 (a) to the others, pore distribution increases as physical mixing ratio is higher. Pores development in HTC180-08-1:1 in Figure 1 (b) which is the highest KOH physical mixing concentration, shows larger pores with estimated size of 5-10μm, surrounded by dispersed smaller pores. Less significant pore structures can be found in HTC180-08-1:2 in Figure 1 (c), as activation concentration is lower than aforementioned sample. And in HTC-180-08-1:3 in Figure 1 (d) barely yields pore development but structures are mildly altered due to hydrothermal treatment.

3.2. Surface chemical properties of prepared porous carbon from nipa palm husks

Fourier transform infrared spectroscopy was used to observe effects of hydrothermal carbonization and chemical activation towards surface chemical behaviors and checkup for activating agent residue.

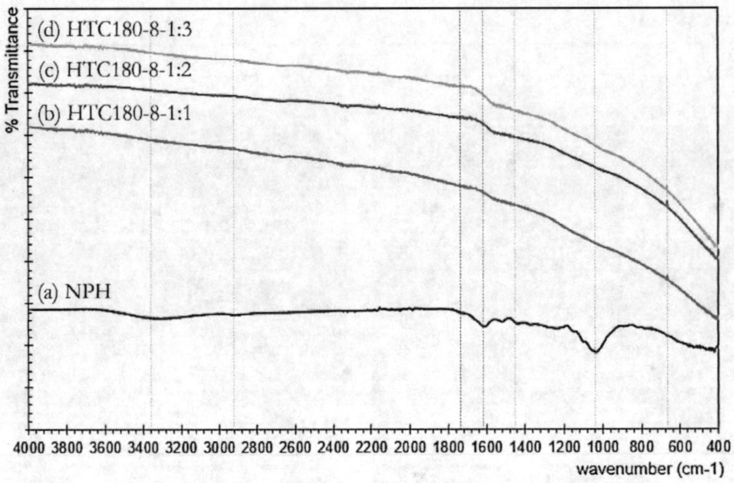

Figure 2. FT-IR spectra of (a) raw nipa palm husk powder, (b) HTC180-08-1:1, (c) HTC180-08-1:2 and (d) HTC180-08-1:3. Vertical lines indicate significant peaks.

According to FT-IR spectra of prepared porous carbon materials illustrated in Figure 2, shows that hydrothermal carbonization helps in restructuring chemical components in nipa palm husk powder by getting rid of organic compounds consist of hydroxyl group at wavenumber of around 2920cm^{-1}, alkyl group at around 1440cm^{-1} and ester/ether group at around 1050cm^{-1}. Moreover, carbonyl group at around 1700cm^{-1} and aromatic C-H bonding at 680cm^{-1} are raised as well. From these transmittance peaks changes from plenty of functional groups in NPH in Figure 2 (a) to carbon materials in Figure 2 (b-d) , can be summed up that aromatic structures are promoted during hydrothermal carbonization with KOH physical mixing, whereas other functional groups in biomass complex structures are almost completely transformed, these affect to enhancing in carbon content of the materials.

3.3. Crystallinity of prepared porous carbon from nipa palm husks

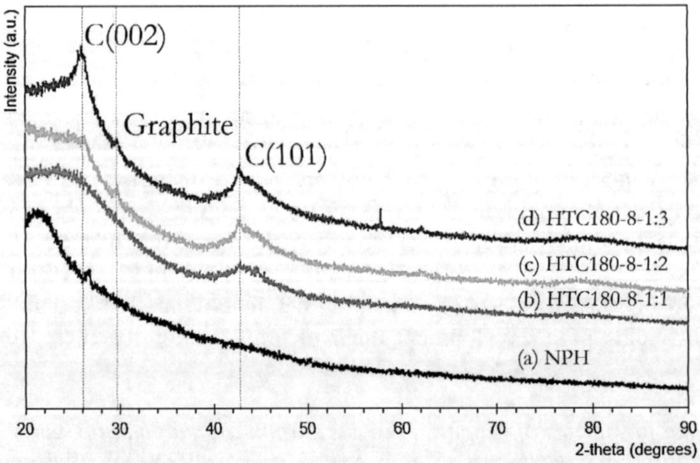

Figure 3 X-ray diffractogram of (a) raw nipa palm husk powder, (b) HTC180-08-1:1, (c) HTC180-08-1:2 and (d) HTC180-08-1:3. Vertical lines indicate significant peaks.

X-ray diffraction technique was used to ensure carbon structures in prepared porous carbon materials and diffractograms are shown in Figure 3. By comparing raw NPH to carbon materials prepared at different mixing ratios. There are new diffraction peaks appeared in carbon materials. Peaks at 2-theta of 26° and 43° indicate carbon material's (002) and (101) planar structures, respectively. Smaller peak near 2-theta of 30° shows graphitic structure.

From the diffractograms, can be clearly pointed out that hydrothermal carbonization and chemical activation also induct carbon restructure. Restructuring occurs by fragmenting organic polymers in biomass by hydrolysis and decarboxylation.[11] During hydrothermal treatment, fragmentation and reconstruction occurs simultaneously, results in increased orderly structures.[12] Moreover, orderly structures tend to be appeared at less severe conditions as in HTC180-08-1:2 and 1:3 has higher intensity at graphite peak than in 1:1.

4. Conclusions

By combining hydrothermal carbonization method with physical mixing-based chemical activation, highly porous carbon material can be prepared using less energy and activation cost. Physical mixing concentration ratio plays a crucial role in pore development both physical and chemical ways. As shown in results, porosity is significantly increased along with higher concentration. For surface chemical aspect, changing activation concentration doesn't alter chemical properties in carbon materials. That means any concentrations can offer satisfied chemical structure transformation. Crystallinity is affected by severity of hydrothermal and activation conditions; milder conditions yields more orderly-constructed carbon structure than in severer ones. From all of these, can be clearly summed up that using a moderate temperature and balanced chemical activation factors are preferred choices to synthesize highly porous carbon materials from biomasses. And exactly, further studies on the most suitable concentrations on each type of biomasses may be required to achieve the most satisfied results.

5. References

[1] Eduardo C, Karla A and Almeida R 2016 *Ciência Rural* **46** pp 1872-77
[2] Minowa T, Kondo T and Sudirjo T 1998 *Biomass Bioenerg.* **14** pp 517-24
[3] Kumar S, Loganathan V A, Gupta R B and Barnett M O 2011 *J. Environ. Manage.* **92** pp 2504-12
[4] Aygün A, Yenisoy-Karakaş and Duman I 2003 *Micropor. Mesopor. Mat.* **66** pp 189-95
[5] Fernandez M E, Ledesma B, Román S, Bonelli P R and Cokierman A L 2015 *Bioresource Technol.* **183** pp 221-8
[6] Azharul Islam Md, Ahmed M J, Khanday W A, Asif M and Hameed B H 2017 *J. Environ. Manage.* **203** pp 237-44
[7] Soares S 2016 *Bol. Grupo Español Carbón* **40** pp 36-41
[8] Erenturk S and Korkut Ö 2018 *Renew. Energ.* **117** pp 374-9
[9] Isikgor F H and Becer C R 2015 *Polym. Technol.* **6** pp 4497-559
[10] Kim Y W, Oh J I, Lee S S, Lee K H, Lee J Ch and Kwon E E 2019 *J. Clean. Prod.* **236** n.p.
[11] Funke A and Ziegler F 2010 *Biofuels, Bioprod. Bioref.* **4** pp 160-77
[12] Sabzoi N, Griffin G and Mujawar M N 2017 *Renew. Sust. Energ. Rev.* **73** pp 1289-99
[13] Sabio M M and Reinoso F R 2004 *Colloid. Surface. A.* **241** pp 15-25
[14] Nowicki P, Kuszyńska I, Przepiórski J and Pietrzak R 2013 *Cent. Eur. J. Chem.* **11** pp 78-85

Acknowledgements

This work was financially supported by Youth Scientist and Technologist Programme (YSTP), National Science and Technology Development Agency (NSTDA), Thailand.

IOP Conf. Series: Materials Science and Engineering **893** (2020) 012003 doi:10.1088/1757-899X/893/1/012003

Preliminary Research of Synthesizing Battery Grade Nickel Sulphate from Dissolution of Nickel Matte in Dilute Sulphuric Acid

Sih Wuri Andayani*, Alfiz Muhammad Qizwini, Harry Setyo Wibowo and Muhamad Arsyansyah

Centre for Material and technical product, Ministry of Industry Republic of Indonesia

* Corresponding author's email: andayani@kemenperin.go.id; wurisih@yahoo.com

Abstract. Nickel matte is a smelter product of Indonesia mining, and it was produced from laterit ores by pyrometallurgy. The nickel matte composition is about 77% Ni, 1% Co and 0.3% Fe. This study is involving with synthesizing nickel sulphate by dissolution of nickel matte in sulphuric acid. The dissolution condition was varied by sulphuric acid grade, temperature and time of process. The dissolution process will be more effective in higher temperature and longer process time, with additional mangan oxide as catalyst. The efficiency process was observed from the filtrate, residue and nickel sulphate. The filtrate was evaluated by using Atomic Absorption Spectrometer, and residue/nickel sulphate were analysed by XRD. The more efficient process of nickel dissolution was observed at 95 °C for 3 hours, with the conversion rate about 55% - 60%. There was no difference result between sulphuric acid technical grade and pure analytic grade. After dissolution process, the synthesizing was using co-precipitation method in varied pH and alcohol/oxalic acid. The characterization of nickel suphate has been carried out by X-ray Diffraction. The most effective of synthesizing of nickel sulphate is adjusting the pH above 5. The product of nickel sulphate will be used a precursor material for lithium battery cathode materials.

1. Introduction

Indonesia is third nickel producer in the world, with the production about 97 Metric tonnages and 69 Metric tonnages as reserved [1]. The nickel can be found in Kalimantan, Sulawesi, Halmahera and Papua Island. The nickel sources in Indonesia is nickel laterite, about 0.8% - 2% of nickel content. The nickel laterite is processed by pyrometalurgy process for resulting either ferro nickel [2] or nickel matte [3]. There are different composition between nickel matte and ferronickel. The nickel matte composition is 77% Nickel, 1% Cobalt, 0.3% Ferrous and 22% Sulfur. And the ferronickel is 17% - 24% Nickel, 1.2% Cobalt and about 70% of Ferrous. They are used as raw materials from certain industries, such as coating, electronic, battery, and etc.

Nowadays, nickel takes an important role for electrical vehicle. Nickel will be used as raw material of lithium battery, as cathode active material. In electric vehicle, lithium battery is functioning as energy storage, with the cost more than half of vehicle's price [4,5]. The lithium battery's main components are cathode, anode, electrolyte and separator. And among of its components, the cathode is the most expensive, and takes about 53% of the total cost [6]. For cathode, there are active materials such as lithium mangan oxide (LMO), lithium ferrophopshate (LFP), lithium Nickel Cobalt Alumunium oxide (Li-NCA), Lithium Nickel Manganese Cobalt oxide (Li-NMC), and etc. Every active material has its own characteristic and costs [7]. Based on the characteristic and cost, many developments have been made to improve its properties. And for Li-NMC, the developments have been made in ratio/composition as well as other feature. The initial combination of Li-NMC was ratio nickel manganese and cobalt 1:1:1. Nowadays, there is a trend in increasing the nickel content and lowering the cobalt content for cathode active material [8,9]. The other ratio is nickel manganese cobalt 5:3:2, and called NMC 532. There others development of ratio of Nickel manganese cobalt, such as 6:2:2 (NMC 622) and 8:1:1 (NMC 811). Their aims are for improving the spesific energy, energy density, cycle life and thermal stability and also cost reduction [10].

Based on the requirement of NMC as the cathode active material, nickel matte could be used as an alternative route for synthesizing cathode active material of Li-NMC [11]. Based on nickel matte composition, it will be more efficient that ferro nickel for synthesising cathode active material. The lower content of ferrous was considered.

There many route for synthesizing of cathode active material. One of the routes is nickel sulphate. The nickel sulphate, combined with cobalt sulphate and manganese sulphate, could be processed to become Li-NMC by using co-precipitation process. This preliminary research was dealing with synthesizing nickel sulphate from nickel matte, as precursor of cathode active material. The nickel matte was dissolved in dilute sulphuric acid in specific condition resulting mixture solution of nickel, cobalt and mangan sulfat [12]. The impurities such as copper and ferrous were eliminated. Copper was eliminated from solution by adding more nickel matte. It was cemented by nickel matte. And ferrous was eliminated by adjusting the pH, and it will precipitate from solution [13]. And synthesizing of nickel sulphate was using co-precipitation method in ethanol and oxalic acid condition [12]. Then, it will be processed with other precursor material such as Cobalt sulphate and mangan sulphate for synthesizing the NMC by co-precipitation method. The co-precipitation method is one of method for producing NMC cathode active material [14, 15].

2. Experimental process

The nickel matte was characterized using electrogravimetry and ICP method. It was prepared by grinding into 200 mesh. The dissolution process was involving dilute sulphuric acid. The residue of solution was separated, and the filtrate was adjusted the pH using ammonium hydroxide. There were variations of time and temperature process. Every variation was evaluated from the mineral content of filtrate and the residue. Also, the effectivity of dilution process was evaluated using XRD Bruker D8 Advance.

The nickel sulphate was synthesized by co-precipitation method in varied of pH and chemical addition. The nickel sulphate as a product was characterized using XRD.

3. Results and discussion

The nickel matte is a smelter product that contains about 70-75% of nickel. The composition of nickel summarized in Table 1. The composition was analyzed using eletrogravimetry and ICP. The data was obtained from process technology analytical laboratory PT. Vale Indonesia Tbk.

Table 1. Chemical composition of nickel matte

	Composition (% w/w)							
	NiCoCu	NiCo	Ni	Fe	Co	Cu	S	P
Nickel matte	78.022	77.879	76.898	0.301	0.98	0.144	21.67	< 0.0001

The nickel matte was dissolved using manganese oxide as catalyst. The result of dissolution process was the mixture of Nickel Mangan Cobalt Sulphate in specific concentration/composition. Based on other research data, the dissolution process of nickel matte was affected by air flow rate, stirring speed, ratio between nickel matte and sulphuric acid, sulphuric acid concentration, process temperature, concentration of oxidator (ferrous sulphate) and particle size [12]. And this research, the process was analyzed from variation of sulphuric acid concentration, temperature and time of process, and also oxidator mangan oxide. The ferrous sulphate could not be used in this result, due to its content should be eliminated. The effectivity of the process was evaluated using AAS and XRD of filtrate and residue. The filtrate was analyzed using AAS that shown in Table 2. Based on Table 2, the more efficient process was involving mangan oxide at temperature process of 95 °C for 3 hours. By increasing temperature of process and manganese oxide addition, there was 5% increasing of Ni dissolution efficiency.

Table 2. The efficiency of dissolution process

Proccess	**Composition (% w/w of nickel matte)**					**Remarks**
	Ni	**Mn**	**Co**	**Fe**	**Cu**	
1	40.26	0	0.21	0.16	0	Without catalyst
2	45.66	0	0.19	0.12	0.07	Without catalyst
3	55.05	13.74	0.24	4.33	0.05	
4	57.50	0.02	0.24	0.16	0.03	T process 95 °C
5	53.27	2.88	0.39	1.75	0.01	

Other than from the filtrate, the efficiency of dissolution process was evaluated from the residue. The residue was analyzing by XRD (shown by Figure 1 and Figure 2). Based on Figure 1 and Figure 2, the dissolution process was not effective. It was observed that there is some nickel matte in the residue. The analysis in both of qualitative and quantitative. The Figure 3 and Figure 4 showed the quantitative analysis. The nickel matte was evaluated as a reference.

In this research, the dissolution process will be more effective in addition of mangan oxide as oxidator, and temperature at 95 °C. Eventhough, the process was ineffective by observing the residue analysis. Based on XRD analysis that shown by Figure 3 – 4 and summarized in Table 3, the nickel matte was dissolved about 53%. Based on the literature [12], the dissolution process should be longer to get more nickel in solution. So, the next research will be done longer, within 36 hours.

From the data of Tabel 3, the residue contains higher nickel than nickel matte. Thus, the dissolution process could reduce nickel matte (Ni (+2)) become nickel (Ni (0)). So, The nickel could be more processed become nickel sulphate by adding more sulphuric acid. And for the next research project, the sulphuric acid will be added more in quantity or higher concentration.

Besides nickel, there is nickel sulphate in the residue, about 33.86%. This is caused by improper separation process. The next process, the filtration process should be improved and developed.

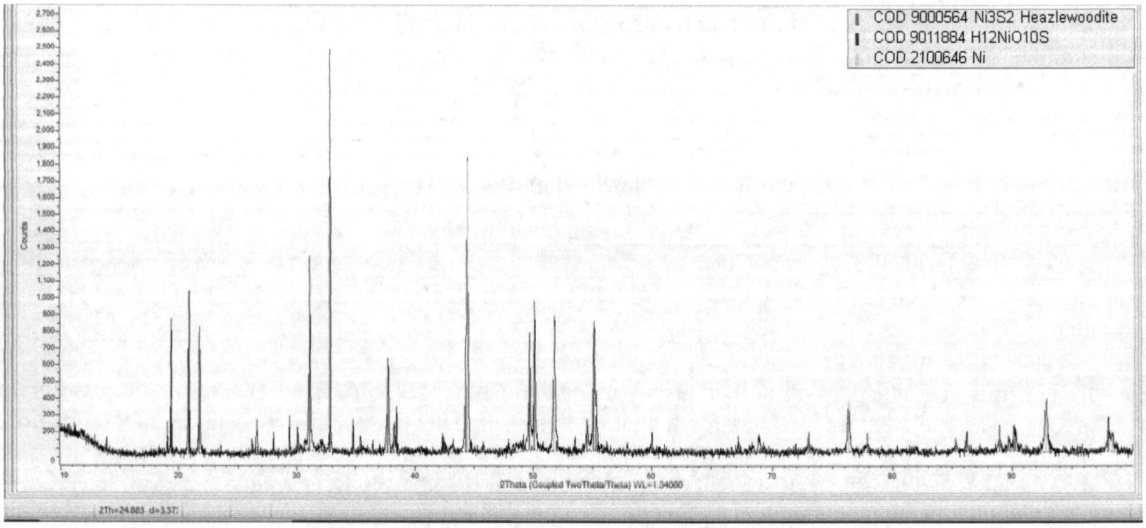

Figure 1. The XRD diffraction of nickel matte

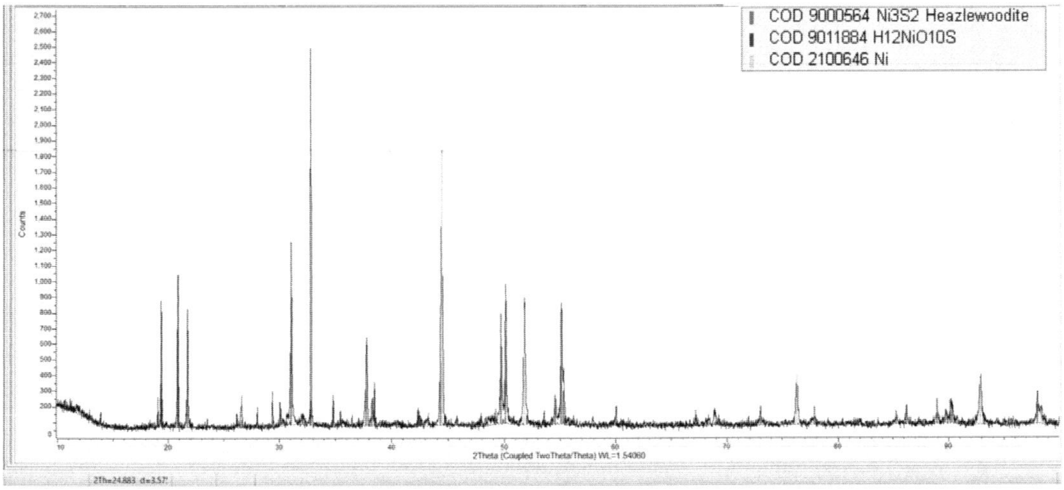

Figure 2. The XRD diffraction of residue

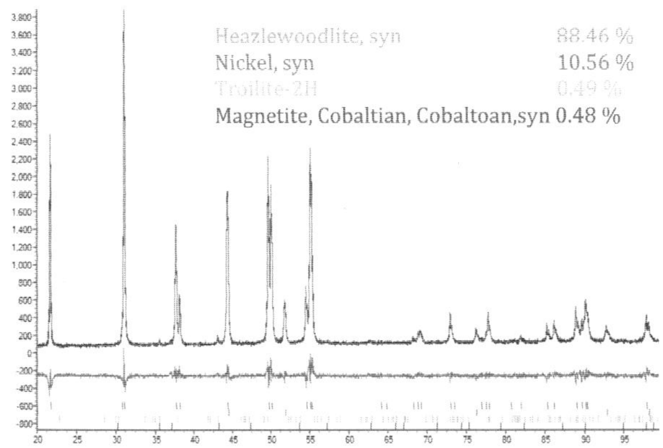

Figure 3. The quantitative analysis of nickel matte

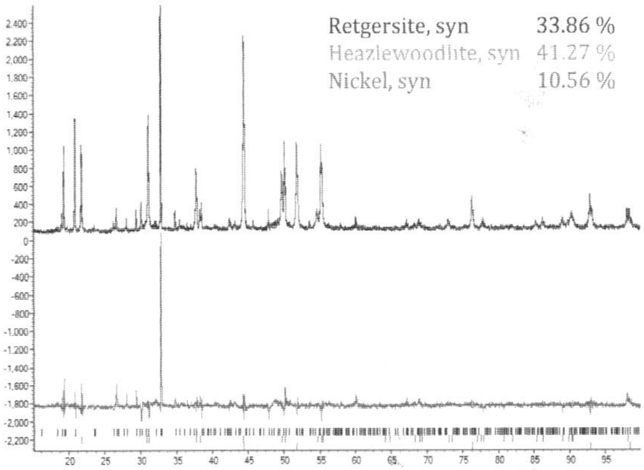

Figure 4. The quantitative analysis of residue

Table 3. The quantitative analysis of dissolution process

No.	Mineral	composition (%)	
		Nickel matte	Residue
1	Haezlewoodites (Ni_3S_2)	88.46	41.27
2	Nickel	10	24.87
3	Retgersite ($NiSO_4$. 6 H_2O)	-	33.86

After dissolution, the process was followed by co-precipitation to produce nickel sulphate. The co-precipitation process was varied from pH and alcohol/oxalic acid. The variations of co-precipitation are pH 1 and pH 5, and addition of alcohol and oxalic acid. The aim of increased pH was eliminating ferrous and solidifying. The effectivity of each process was evaluated of nickel sulphate using XRD. The diffractions were shown by Figure 5-Figure 8.

Figure 5 and 6 showed the nickel sulphate was impured by ammonia. The ammonia was added to increase the pH 1 - 5. Thus, the next research project should be developed the washing process.

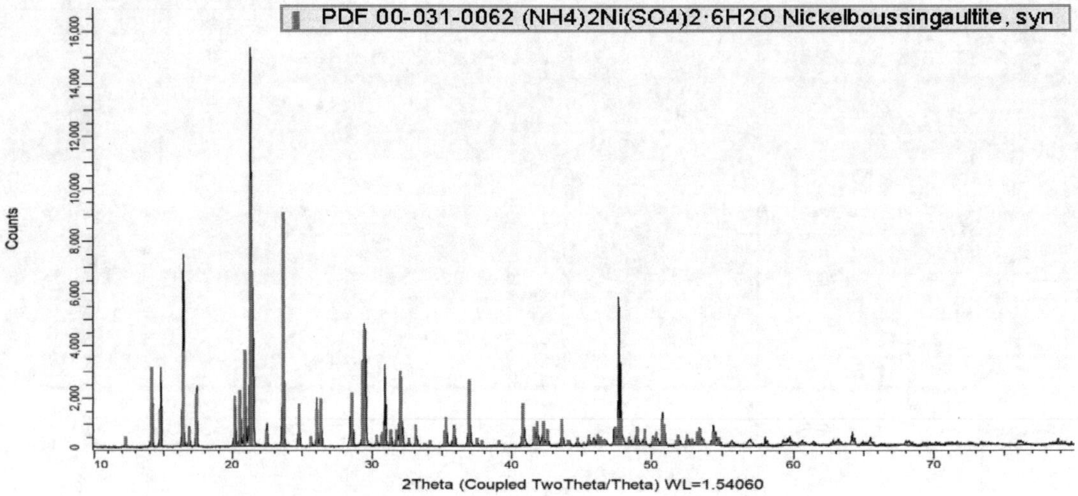

Figure 5. The XRD diffraction of nickel sulphate at pH 1 process

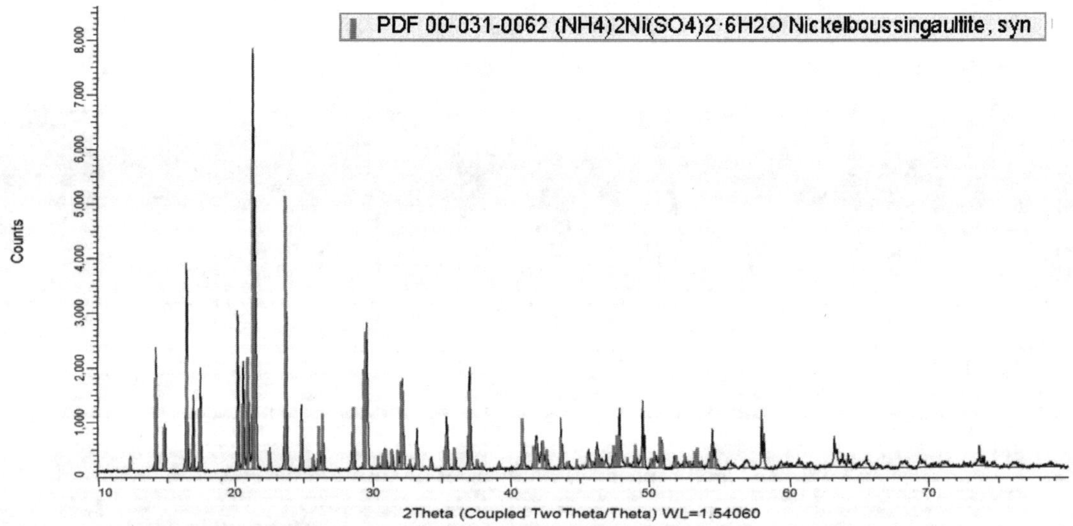

Figure 6. The XRD diffraction of nickel sulphate at pH 5 process

The pH increasing was aimining for co-precipitated the nickel sulphate, also for eliminating of ferrous (Fe). The increasing of pH did not eliminated Fe effectively. There was 21% higher of Ferrous removal in higher pH. So, The process should be developed for better eliminating Fe process.

Besides adjusting the pH, the co-precipitation process for synthesizing of nickel sulphate was achieved by alcohol and oxalic acid addition. The effectivity of co-precipitation was evaluated by XRD, as shown by Figure 7 and Figure 8.

Based on Figure 7, the co-precipitation using oxalic acid did not result nickel sulphate, but nickel oxalic. So, this process will not be further developed. Eventough, the co-precipitation by alcohol resulted nickel sulphate, with impurities of potassium. It should be analysed more detail.

By observing the diffraction of nickel sulphate, the most co-precipitation effective method was adjusting the pH above of 5. But the process should be developed for resulting nickel sulphate as precursor material of active material for cathode (battery grade).

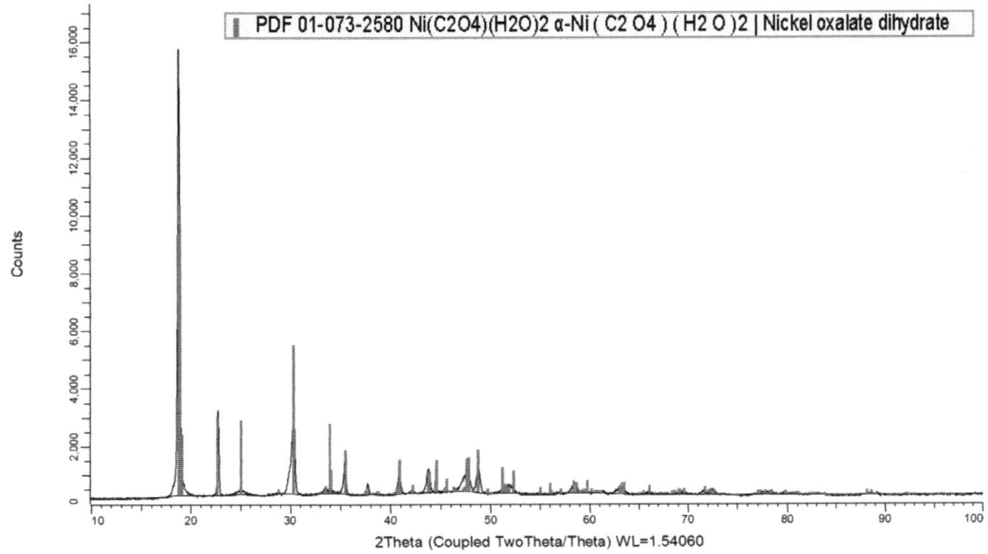

Figure 7. The result of co-precipitation by oxalic acid

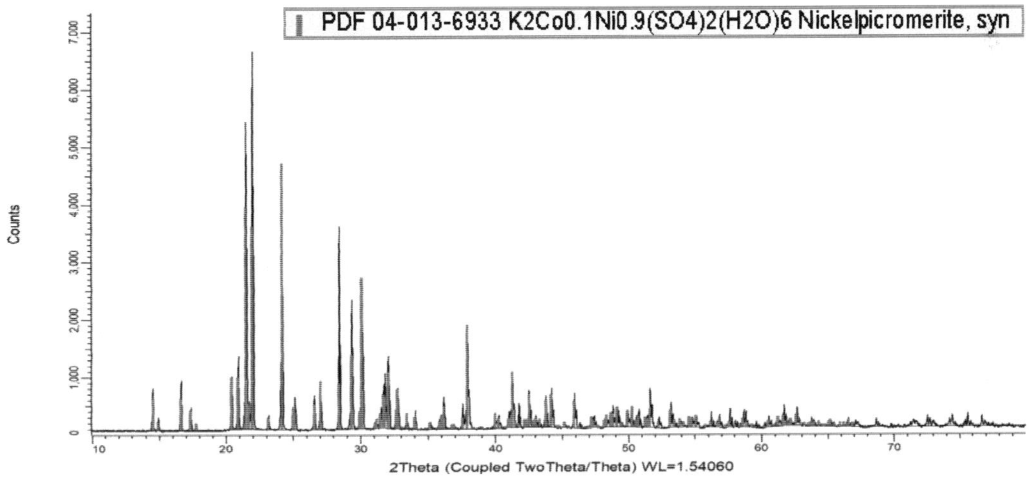

Figure 8. The result of co-precipitation by alcohol

4. Conclusion
The preliminary study of synthesizing of battery grade nickel sulphate could be evaluated by analysizing of residue and filtrate of dissolution process of nickel matte, and the quality of nickel sulphate. The following conclusion had been achieved:

1. The nickel sulphate could be synthesized from nickel matte by dissolution process in dilute sulphuric acid and continued by co-precipitation method by adjusting the pH above of 5. The purity of nickel sulphate should be improved to eliminate the ammonia.
2. The nickel sulphate could be used as a route for synthesizing cathode active material of lithium battery.
3. The research should be develeped about dissolution process time, additional oxidator and catalyst, co-precipitation process and washing and separation process, etc. Also, there should be comparative study between nickel sulphate and battery grade nickel sulphate commercial.

5. Reference
[1] www. psdg.geologi.esdm.go.id

[2] www.antam.com
[3] www.vale.com
[4] www.otomotif.bisnis.com
[5] www.easyelectriclife.groupe.renault.com
[6] www.qnovo.com
[7] www.thedeepdive.ca
[8] Lucas Weimer, Tobias Braun, and Ansgar vom Hemdt 2019 *Resources policy* **64** 101473
[9] Marc wentker, Matthew Greenwood and Jens Leker 2019 *Energies* **12** 504
[10] www.batteryuniversity.com
[11] Zhao Xi, Zhixing Wang, Guochun Yan, Huajun Guo, Xinhai Li, Qiyang Hu, Wenjie Peng, Jiexi Wang 2019 *Hydrometallurgy* **186** 30
[12] Mehmet Hakan Morcali, Leili Tafaghodi Khajavi, Serdas Aktas, David B. Dreisinger 2019 *Hydrometallurgy* **185** 257
[13] Rudi Subagja 2018 Proc. Int. Conf. on Mineral Processing and Technology (Jakarta) vol 285 (UK: IOP) p 012001
[14] Qiang Dai, Jarod C. Kelly, Jennifer Dunn, and Pahola Thatiana Benavides 2018 *Update of Bill – of – Materials and Cathode Materials Production for Lithium – ion Batteries in The GREET Model* (Chicago: Argonne National Laboratory – Energy System Divisions)
[15] Shaber Ahmed, Paul A. Nelson, Kevin G. Gallagher, Naresh Susarla, and Dennis W. Dees 2017 *Journal of Power Sources* **342** 733

Acknoledgement
The research of preliminary study of battery grade nickel sulphate from nickel matte was funding by research grant of Centre for Material and Technical Product Ministry of Industry. The author thanks to PT. Vale Indonesia Tbk and PT. Weda Bay for nickel matte and nickel hydroxy carbonate supply.

IOP Conf. Series: Materials Science and Engineering **893** (2020) 012004 doi:10.1088/1757-899X/893/1/012004

Evaluation of Noise Absorption Capacity of Various Materials

C Singh[1], M Kumar[1] and M L Chandravanshi[1]

[1] Department of Mechanical Engineering, Indian Institute of Technology (Indian School of Mines), Dhanbad, India

Abstract. Noise absorption and transmission property of materials is of great importance in order to determine the amount of noise transmitted from source to receiver. Unwanted noise can cause various biological and psychological effects to humans and other organisms in nature. This paper deals with the experimental study on noise absorbing and deflecting ability of various available materials. These materials could be utilized in industrial complexes and building construction applications for noise level reduction purposes. The materials such as glass, plywood, thermocol (scientifically known as polystyrene), fiber plastic sheet and coconut fiber sheet have been used in the experiment as a noise barrier. Through experimental analysis, the reduction noise level was obtained in terms of decibel (dB), sound power level and loss of sound energy using hand held noise analyzer. The finding suggests that thermocol has the best sound absorbing capacity at medium to low sound pressure levels, whereas glass has the best sound absorbing capacity at high sound pressure levels.

1. Introduction

Noise is generally defined as an unwanted or unpleasant sound which causes undesired effects. Various methods have been adopted in the last few decades to reduce the noise levels for increased pleasure and comfort. One of the major contributions to this cause has been the use of sound absorbing materials. Various materials with different sound absorbing capacities have been used effectively. Mihai Bratu et al. [1] presented a research on the sound absorbing capacity of composites reinforced with various types of wastes which are harmful to the environment. The authors determined the absorption coefficient for each sample and deduced the effect of reinforcement material on the absorption coefficient. Md. Ayub et al. [2] analyzed the sound absorption capacity of natural coir fiber using Delany-Bazley model. The analytical results were validated using experimental investigation. It was concluded that the proposed method perfectly justifies the overall trend of absorption coefficient. Shu yang et al. [3] investigated the sound absorption properties of fiber assemblies using standing wave tube method. The authors also proposed a theoretical model to analyze and predict the absorption coefficient as a function of frequency. Similar analysis were done on sound absorption behavior of knitted fabrics by Yanping Liu and Hong Hu [4] and on asphalt mixtures by Ramon Mendes Knabben et al. [5]

Similar research work has been conducted by other researchers on woven fabric [6, 7] and knitted fabric [8, 9]. Research work has also been conducted on porous polymer spheres [10] and perforated panels [11, 12]

The literatures have suggested different methods to analyze the noise absorption capacity of the material. The objective of this paper is to estimate the noise data in the field regarding reduction in noise levels using various easily available materials. The study may be useful for various apartments in the densely populated cities such as in India to reduce the noise levels in terms of sound energy and noise levels (dB*(A)*) scale for each materials individually. Further, mathematical analysis has also been done to calculate the power transmitted and energy loss through the noise barrier for different materials.

1.1 Mathematical background

It well known that, whenever a sound travels from one medium to another medium, some fraction of sound is reflected from the surface, some fraction is absorbed by the medium and the remaining part is transmitted from one side of the barrier to other. The sound energy lost during this process is called the transmission loss, and is given by the equation-

$$Transmission\ Loss\ (TL) = IncidentPower\ (IP) - TransmittedPower\ (TP) \tag{1}$$

Incident power is given by the equation-

$$IP = 10^{L_i/10} * P_{ref} \tag{2}$$

Where, L_i is the incident sound level in dB, $P_{ref}$is the reference is the reference sound power (10^{-12} Watts)

Similarly, transmitted power is given by-

$$TP = 10^{L_t/10} * P_{ref} \tag{3}$$

Where, L_t is the transmitted power in dB,

The reduction in sound energy when passing through a medium is given by the equation-

$$\Delta E = 10^{\frac{TL}{10}} \tag{4}$$

Where, $\boldsymbol{\Delta}$E is the reduction in sound energy and TL is the total transmission loss

During the transmission, some amount of energy is lost through leakage. The power transmission taking the leakage into consideration is given by the equation-

$$P_t = IP * Attemuation$$

Or

$$P_t = IP * (1 - LF) * \frac{1}{TL} \tag{5}$$

Where, P_t is the power transmitted, LF is the leakage factor (LF has been considered as 1% in this article.

2. Experimental setup

The experiments were conducted by preparing a scale down model of a house using bricks and clay cement. The dimensions of the closed cavity are 50x30x40 cm^3. The instrument used for noise measurement was type 2270 hand held noise analyzer by Bruel and Kjaer. The experimental setup and the instrument used have been shown in Figure 1.

IOP Conf. Series: Materials Science and Engineering **893** (2020) 012004 doi:10.1088/1757-899X/893/1/012004

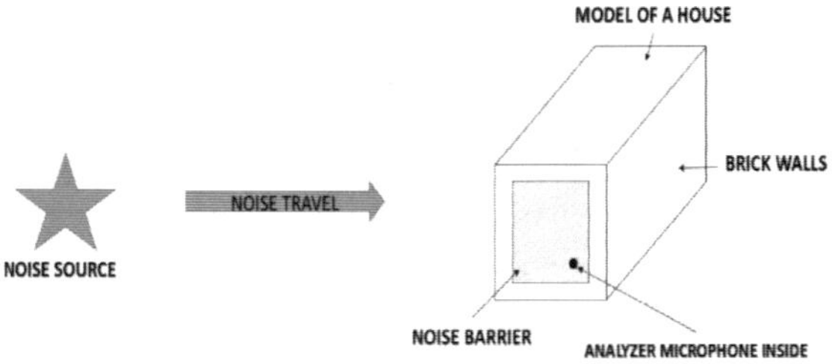

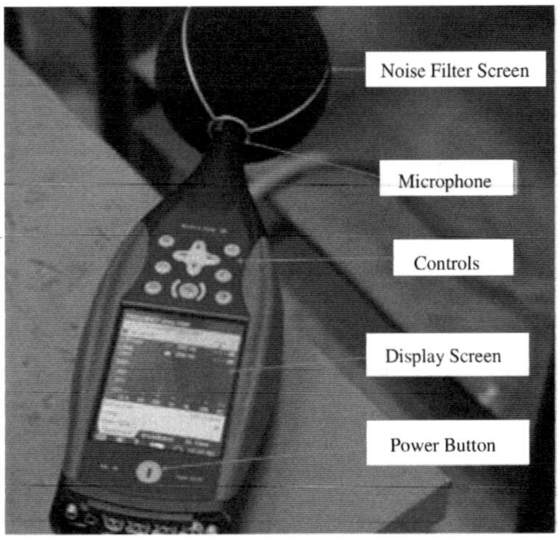

Figure 1. (a) Closed cavity (model) developed for experimental setup; (b) Bruel and Kjaer type 2270 hand held analyzer

One of the faces of the model was kept open to fix the noise barrier material to that face. To take the observations, different materials were fixed on the open face one by one. A motor vehicle engine was used as the noise source. Different noise levels were generated by varying the RPM of the engine. A noise analyzer was placed inside the box model and the noise barriers were used to seal the model. The materials used as noise barrier in the experiment were:- glass, plywood, thermocol, plastic fiber and coconut fiber sheets as depicted in the Table 1.

Table 1. Noise barriers used for the experimental data analysis

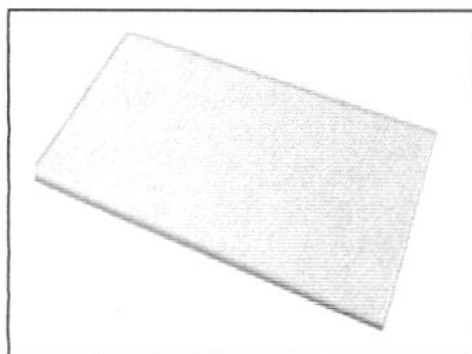	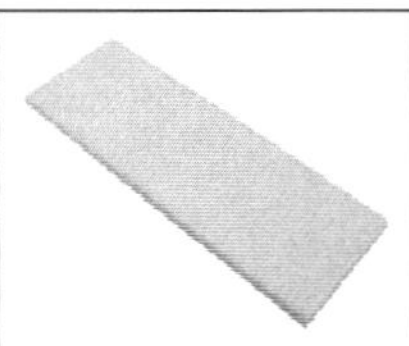			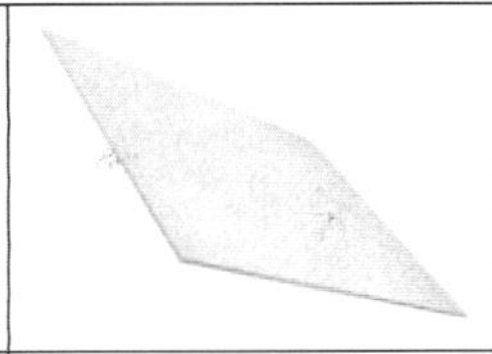
Glass Sheet	Themocol sheet	Plywood sheet	Coconut fiber sheet	Plastic fiber sheer

For data acquisition purpose, the noise analyzer and microphone were connected through extended cables. The microphone was kept inside the especially cavity with mortars and bricks. The motorcycle engine speed was varied sequentially in order of 2000 rpm up-to 10000 rpm and noise levels were recorded in dB(A) i.e. decibel scale with *A-weighing.* The dB(A) scale has been selected as it represents the human ear response to sound. Ambient condition readings were also taken without any noise barrier. The recorded observations were then formulated using equations (1) to (5) and the comparative analyses were carried out to determine the materials with best sound absorption capacity.

3. Results and discussion

Figure 2 shows the comparative study of equivalent noise levels inside the room in an ambient condition and the active mode noise source for estimating the noise transmitted to the room with each of the materials as noise barrier at various engine speeds individually.

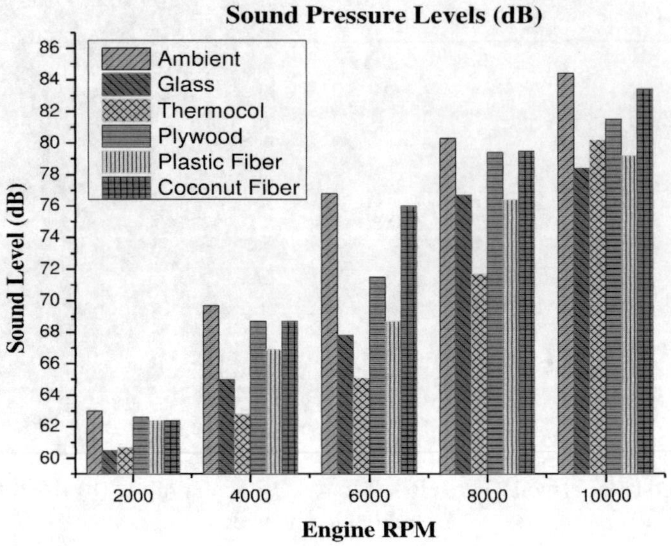

Figure 2. Noise pressure level (dB) with different material noise barrier

The comparative study of noise levels as shown in Figure 2 revealed that the noise absorbing capacity is the higher for thermocol and glass sheets when compared to the plastic fiber, coconut fiber and plywood sheets.

Another parameter used for assessing the noise absorption property of the material is the pressure transmission ratio. It is the ratio of pressure incident on materials to the pressure transmitted from the materials. These values of transmission ratio have been obtained from experimental data observations and the average values have been depicted in Figure 3.

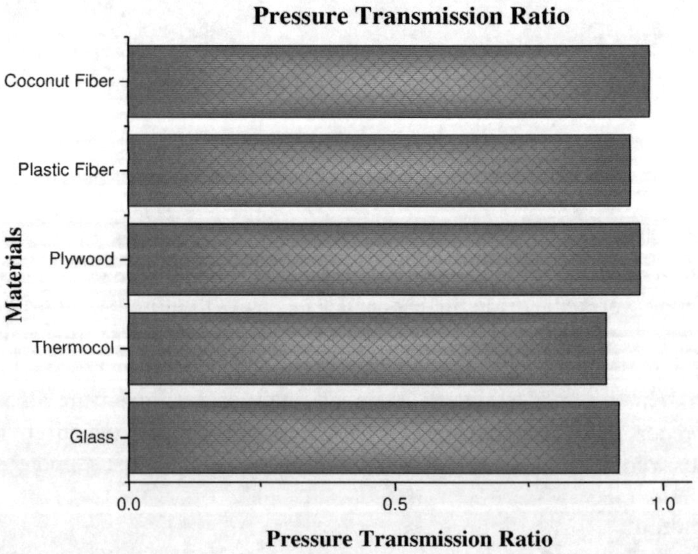

Figure 3. Pressure transmission ratio of different materials

The above figure depicts that thermocol has the lowest pressure transmission ratio i.e. thermocol has the best sound absorption capacity followed by glass and plastic fiber sheet. The sound absorption capacity of thermocol is due to its porosity which helps to damp the noise. Coconut fiber has the highest transmission ratio. It is due to the fact that the coconut fibers have micro-voids in them which allow noise to pass through.

In order to understand in depth about noise absorbing capacity of each material, the reduction in sound power and sound energy levels have been calculated using the formulas 1 to 5 and presented in Figure 4 (a) and (b) respectively.

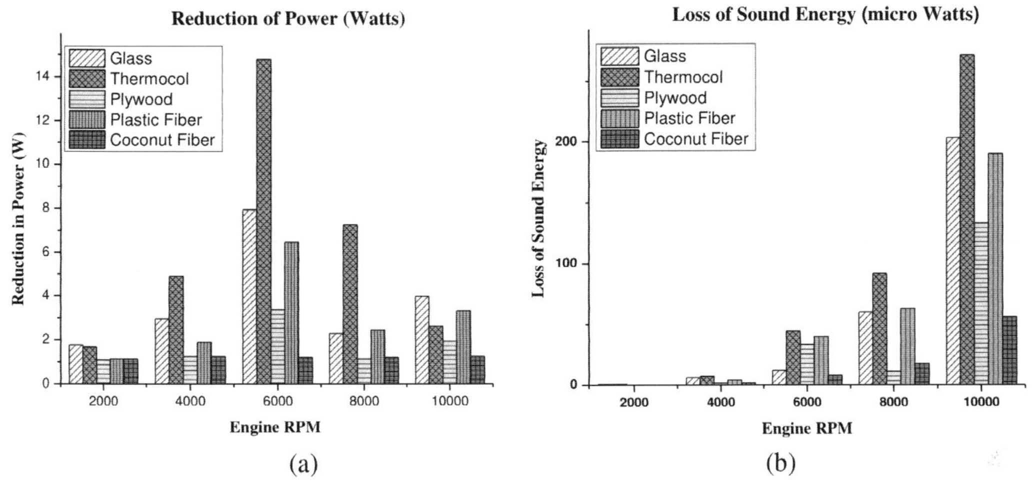

Figure 4. (a) Reduction in sound power for each material; (b) Loss of sound energy for each material

Figure 4 indicates that the reduction of sound power and loss of energy is maximum for thermocol barrier followed by glass barrier. Plastic fiber sheet shows the potential to be developed in composite form to be used as a replacement to heavy glass sheets. Another important observation is that although the noise absorption of coconut fiber is low, but it is the most flexible, cheap and easily available material from the enlisted materials. Also, coconut fiber can be used in composite form to improve the noise absorption capacity and provide better results.

4. Conclusion

The study was focused at determining the noise absorbing capacity of various materials. The experiments were performed to record the sound pressure levels and calculations were done to determine the reduction in power and loss of sound energy through each material. The experimental results and the mathematical analysis suggested that the noise absorbing capacity of thermocol and glass sheets was the highest among the considered materials. It was also observed that glass provides better noise damping at higher noise levels whereas thermocol provides better damping at low to medium noise levels. Another interesting observation was the noise reduction due to plastic fiber sheet. The plastic sheet had results comparable to that of glass sheet. Hence it can be said that instead of using heavy glass sheets, a plastic fiber sheet and thermocol can be manufactured in a composite form as a light weight and cheap material which is capable of providing high noise damping. It can also be said that the flexibility and easy availability of coconut fiber can be mixed with porous materials and can be used to develop composites with high noise absorption capacities.

5. References

[1] Bratu M, Ropota I, Vasile O, Dumitrescu O and Muntean M, (2011) Sound-Absorbing Properties of Composite Materials Reinforced with Various Wastes, *Environmental Engineering and Management Journal,* **10**(8): 1047-1051.

[2] Md. Ayub, M. J. Mohd Nor, Nowshad Amin, Rozli Zulkifli, M. HosseiniFouladi and A. Rasdan Ismail, (2009) Analysis on Sound Absorption of Natural Coir Fiber using Delany-Bazley Model, *Proceedings of the International Conference on Mechanical Engineering (ICME 2009).*

[3] Shu Yang, Weidong Yu and Ning Pan, (2010) Investigation of the sound-absorbing behavior of fiber assemblies, *Textile Research Journal,***81**(7): 673-682.

[4] Yanping Liu and Hong Hu, (2010) Sound Absorption Behavior of Knitted Spacer Fabrics, *Textile Research Journal,* **80**(18): 1949-1957.

[5] Ramon Mendes Knabben, Glicerio Triches, Samir N.Y. Gerges and Erasmo F. Vergara, (2016) Evaluation of Sound Absorption Capacity of Asphalt Mixtures, *Applied Acoustics,* **114:**266-274.

[6] Shoshani Y and Rosenhouse G, (1990) Noise absorption by woven fabrics, *Applied Acoustics***30**: 321–333.

[7] Shoshani YK, (1991) Noise absorption by a combination of woven and nonwoven fabrics, *Journal of Textile Institute*, **82**: 500–503.

[8] Dias T and Monaragala R, (2006) Sound absorption in knitted structures for interior noise reduction in automobiles, *Measurement Science and Technology,* **17**: 2499–2505.

[9] Dias T, Monaragala R and Lay E, (2007) Analysis of thick spacer fabrics to reduce automobile interior noise, *Measurement Science and Technology*,**18**: 1979–1991.

[10] Hong Z, Bo L and Guangsu H, (2006) Sound absorption behavior of multi-porous hollow polymer micro-spheres, *Materials Letters*,**60**: 3451–3456.

[11] Lee YY, Lee EWM and Ng CF, (2005) Sound absorption of a finite flexible micro-perforated panel backed by an air cavity, *Journal of Sound and Vibration*,**287**: 227–243.

[12] Takahashi D and Tanaka M, (2002) Flexural vibration of perforated plates and porous elastic materials under acoustic loading. Journal of Acoustic Society of America, **112**: 1456–1464.

IOP Conf. Series: Materials Science and Engineering **893** (2020) 012005 doi:10.1088/1757-899X/893/1/012005

Optimization of Cutting Parameters in Machining Polyoxymethylene Using RSM

M Aruna

Department of Mechanical Engineering, College of Engineering and computing, Al Ghurair University, Dubai, UAE

E-mail: maruna9@gmail.com

Abstract. Polyoxymcthylene (POM) is an extensively used engineering thermoplastic polymer. Owing to its superior properties and potential applications in numerous turfs of structural machineries, it is essential to probe the machining of POM. The economy of machining operations plays an imperative role in increasing productivity and effectiveness. The persistence of optimal cutting parameters, such as cutting speed, feed and depth of cut, which are valid for given cutting tools, is one of the dynamic segments in process planning of metal parts. This research work concerns an experimental study dealing with cutting parameters and its effects on surface roughness (Ra) and material removal rate (MRR) during the turning of POM. The approach is based on Response Surface Technology (RSM). Experiments are designed using central composite design and second-order quadratic models are developed to define the optimal machining parameters. These optimized parameters are validated experimentally.

1. Introduction

Polymeric materials have been extensively used to replace traditional metallic materials owing to their high elastic properties. However, polymeric materials can be manufactured as near net shapes, machining is still essential to create the assembly of the products. Due to high ductility and low melting point, the selection of cutting tool and machining conditions plays an important role in machining [1]. In this work, the machining behaviour of high performance engineering polymer, such as ultra-high-molecular-weight polyoxymethylene has been investigated using central composite design. Polyoxymethylene (POM), also branded as Acetal, is an engineering thermoplastic, commonly used in construction machinery, vehicles, technology, transport and cargo handling, electrical, precision technology, domestic, food industry and medical appliances, where very high accuracy is needed in these machined parts [2]. The foremost raw materials are acetal copolymer and homopolymer. POM combines high stiffness with mechanical strength, besides revealing good elastic properties, high toughness, dimensional stability, excellent sliding friction characteristics and high hardness. Furthermore, POM has good fatigue properties and machinability. The beneficial assets of POM comprises Rockwell hardness of 122 (6 % more than nylon), Thermal expansion coefficient 0.000047 in /'F (14 % less than nylon), moisture absorption of 0.9% (87 % less than nylon).

Numerous problems associated with the precision and efficiency in cutting composites has become an essential topic in the manufacturing industry. Authors have reported several investigations on the machining mechanisms and production cost, such as tool wear, selection of machining parameters and control of surface finishing. A significant amount of research activities have also focused on the machinability of a wide variety of reinforced polymeric materials. Some authors [3–5] have compared the machining behaviour of pristine and reinforced polymers based on the characteristics of the matrix phase, which is presented in both materials. However, only a limited amount of article is available to discuss the machining parameters of engineering polymers and their effect on the surface finishing. Investigators examined mainly the machining of phenolic composites [6]. The existence of a critical machining speed was recognized based on the specific cutting pressure and machining temperatures, which directly affect the tool wear. Machinability of polyether-ether-ketone (PEEK) reinforced with glass fibre using polycrystalline diamond (PCD) and cemented carbide (K20) tools are investigated by few researchers [7]. A better surface finishing was achieved when high cutting speed, low feed rate and PCD tool were set.

Orthogonal experiment is conducted to examine the cutting of polyamide (PA6) with and without reinforcing phase (30% glass fibre) using cemented carbide cutting tools. The consequence of glass fibres on the friction angle, shear plane angle, normal and shear stresses and chip deformation in several cutting conditions were assessed. The outcomes directed that the accumulation of glass fibres led to higher cutting force, friction angle and normal stress values [4]. In contrast, the polyamide without reinforcing attained greater shear stress and chip deformation values. Critical thrust forces are predicted and proved the consequence of different types of drills, such as twist drill, saw drill, candlestick drill, core drill and step drill, on the delamination physiognomies [8]. The thermal properties of engineering plastics significantly affected the surface roughness of drilled holes. The investigations on the machinability and hole dimensional accuracy in the drilling process of engineering plastics are left pending. The drilling forces and indemnities in the work piece material were much affected by the drill geometry factor. The viscoelastic properties of engineering plastics affected the cutting force and the surface roughness of machined parts [5]. The polyacetal or polyoxymethylene (POM) and polyetherimide (PEI) materials were drilled using 1 mm diameter drill. The hole accuracy (radius error) befits worse when the drill feed rate is lower and the spindle speed is higher [9].

Plastic parts are produced widely by injection moulding. Precise plastic components can be obtained by machining of plastics [10]. The surface quality of machined plastic components is greatly affected by feed rate and speed. While machining plastics, speed and feed has to be considered as significant factors for attaining dimensional accuracy and good surface finish [11]. Optimal cutting conditions has to be determined to accomplish superior machining performance in machining of polymers [12]. Optimum cutting parameters namely speed, feed rate depth of cut for plastic machining could be identified using regressions analysis and optimum cutting conditions offer better surface finish [13]. Experiments are designed using Taguchi method. This technique includes response table, response graph, normal probability plot, ANOVA. The result shows the rubberized fly ash geopolymer has potential to be machined and can be used for many engineering applications [13]. Response surface methodology is to express the output parameters. Experiments can be designed using Box behnken design plan as they have fewer design points making it less expensive to run than central composite design. Surface models developed for surface shows the effect of each input parameters and its effect with other parameters, depicting the trends of response [14]. Experiments and analysis could be conducted using design of experiments and optimum cutting parameters for machining of POM can be determined by measuring the surface roughness for each treatment combination and main effects and interaction effect can be studied by ANOVA [15].

Abundant research has focused on the material and tribological properties of POM. Determining the optimal process parameters is found to be the major issue in machining Polyoxymcthylene. However, only a limited number of investigations about the effects of the turning and drilling parameters and the machining behaviour of such materials have been reported. The objective of this work is to optimize the process parameters namely cutting speed, feed rate and depth of cut while turning operations is performed on POM. Taguchi's techniques have been used commonly in engineering design, and can be pragmatic to several facets such as optimization, experimental design, sensitivity analysis, parameter estimation, model prediction, etc. Taguchi based optimization technique has produced an exceptional and dominant optimization discipline that varies from traditional practices. Taguchi method practices a distinct highly fractionated factorial designs and other categories of fractional designs achieved from orthogonal arrays (OAs). It is to study the complete experimental region of interest for experimenter with a lesser number of experiments. This diminishes the time and costs of experiments, and furthermore permits for an optimization of the process to be completed. The columns of an OA represent the experimental parameters to be optimized and the rows represent the individual trials (combinations of levels).

Previous research works are limited to ANOVA. In this research work, RSM technique is focussed principally to develop Response surface model. The cutting parameter combinations are designed by Taguchi's Design of experiments (DOE). The surface roughness values and Material removal rate for various parameters are measured and Response surface methodology is used to analyze and validate the results of experimental measurements.

2. Methodology

Response surface methodology is an advanced tool nowadays which involves three factorial designs giving number of input (independent) factors and their corresponding relationship between one or more measured dependent responses. It is advantageous over conventional methods available and it includes less experiment numbers, its suitability for multiple factor experiments and search for common relationship between various factors towards finding the most suitable production conditions and forecast responses. RSM is a collection of mathematical and statistical techniques that are useful for modelling and analysis of problems in which a response of interest is influenced by several variables and the objective is to optimize this response [16]. The version 7 of Design Expert software is used to develop the experimental plan for RSM. The same software is also used to analyze the data collected.

The cutting parameters are to be designed using a response surface design. Experiments are conducted using different levels of cutting parameters. Once the experiments are carried out, the surface finish is measured using surface roughness tester and material removal rate is calculated. In the realistic application of RSM, it is necessary to extend an approximating model for the true response surface. The approximating model is based on the observed data from the process or system and is an empirical model. Surface roughness is presumed to be a task of prime variables cutting speed, feed rate and depth of cut. Multiple regression as a collection of statistical techniques is useful for building the types of empirical models requisite in RSM. The central composite design is used, since it gives a comparatively accurate prediction of all response variable averages. Central Composite Design (CCD) is nearly rotatable second-order design based on three-level incomplete factorial designs. In this design, the treatment combinations are at the midpoints of edges of the process space and at the centre.

RSM designs also assists in quantifying the relationship amongst one or more measured responses and the vital input factors. In order to decide the relationship between the factors and the response variables investigated, the data collected must be investigated in a statistically sound mode by means of regression. Regression is performed in order to describe the data collected whereby an observed, empirical variable (response) is approximated based on a functional relationship between the estimated variable, yest and one or more regressor or input variable $x_1, x_2, \ldots, x_i$. In the case where there exist a non-linear relationship between a particular response and three input variables, a quadratic equation (1),

$$\text{Surface roughness} = -1.94909 + 0.013162\text{Cutting speed} + 45.00778\ \text{Feed} + 2.68761\ \text{Depth of cut}+\ldots \tag{1}$$

$$y_{est} = b_0 + b_1 x_1 + b_2 x_2 + b_3 x_3 + b_4 x_1 x_2 + b_5 x_1 x_3 + b_6 x_2 x_3 + b_7 x_1^2 + b_8 x_2^2 + b_9 x_3^2 + error \tag{2}$$

Equation (2) may be used to describe the functional relationship between the estimated variable, *yest* and the input variables x_1, x_2 and x_3. The least square technique is being used to fit a model equation containing the said regressors or input variables by minimising the residual error measured by the sum of square deviations between the actual and the estimated responses. This involves the calculation of estimates for the regression coefficients.

The present work mainly demonstrates the optimisation process of multiple responses i. e. MRR and Surface roughness using RSM technique. This multi-objective optimization is applied to obtain minimum surface roughness and maximum material removal rate.

IOP Conf. Series: Materials Science and Engineering **893** (2020) 012005 doi:10.1088/1757-899X/893/1/012005

2.1. *Cutting conditions*

In this examination three factors are being studied and their low and high levels are given below.

Cutting speed (m/min):	200 - 260
Feed (mm/rev):	0.1- 0.4
Depth of cut (mm):	0.5 - 2.5

2.2. *Experimental work*

The primary experiments were conducted on the universal lathe machine. The material used for cutting was unreinforced Polyoxymcthylene. Polyoxymcthylene and other thermoplastics can be turned on a lathe with tools ground for plastics, at speed up to 550 m/min. In the present study, three cutting parameters, namely, cutting speed (Vc), feed rate (f) and depth of cut (ap) were considered. The cutting parameter ranges were selected based on machining guidelines provided by workpiece and tool manufacturer's recommendations and previous researches. Three levels for cutting speed, feed rate and depth of cut were considered. The POM rods are machined at different levels of speed, feed and depth of cut, as per the experimental design. MRR is the volume of material removed per minute. The higher you're cutting parameters, the higher the MRR. Surface roughness often reduced to roughness, is a component of surface texture. It is computed by the deviations in the direction of the normal vector of a real surface from its ideal form. If these eccentricities are huge, the surface is rough; if they are small, the surface is smooth. In surface metrology, roughness is normally measured to be the high-frequency, short-wavelength component of a measured surface. However, in practice it is often essential to know both the amplitude and frequency to ensure that a surface is fit for a purpose. There are many different roughness parameters in use, but is by far the most common, though this is often for old reasons and not for particular merit Ra, as the early roughness meters could only measure. Other common parameters include Rq, Rt, and Rz. The surface roughness is measured for the different combinations, using Mitutoyo SJ-410 portable surface roughness tester. It is measured at three different locations and the averages reported. The trials are performed in a random fashion. From the surface roughness value, a regression model is developed. The fit of the model and the significance of the different factors is checked using ANOVA.

2.3. *ANOVA analysis*

F test analysis of variance (ANOVA) is executed and presented in Table 1 and 2, to check the statistical significance of the model. An ANOVA table is commonly used to summarize the tests performed. The analysis of variance (ANOVA) is applied to test adequacy of the developed models. The purpose of the ANOVA is to investigate which parameters significantly affect the response characteristics. The test for significance of the regression model and the test for significance on individual model coefficients are performed. Examination of the fit summary output reveals that the linear model is statistically significant for surface roughness. Table 1 shows the ANOVA table for response surface linear model for surface roughness. It is obvious from the results of ANOVA that the speed rate is the dominant factor affecting surface finish. The contribution of the feed and depth of cut is 9.35 and 1.48 respectively. The Model F-value of 3.62 infers the developed model is significant. P-values <0.0500 specify model relations are significant. Values larger than 0.1000 specify the model terms are not significant. Mathematical model is established using multiple regression method, Ra model correlation R^2 is 93.50%. The Lack of Fit F-value of 0.63 infers the Lack of Fit is not significant comparative to the pure error. There is a 75.42% chance that a Lack of Fit F-value will exist. This large value occurs due to noise. The Predicted R^2 of 0.0935 is in reasonable agreement with the Adjusted R^2 of 0.2929.

Table 1. ANOVA table for surface roughness.

Source	Sum of Squares	df	Mean Square	F-value	p-value	
Model	723.23	3	241.08	3.62	0.0361	significant
A-Cutting Speed	2.13	1	2.13	0.0320	0.8603	

B-Feed	622.46	1	622.46	9.35	0.0075	
C-Depth of cut	98.65	1	98.65	1.48	0.2410	
Residual	1064.66	16	66.54			
Lack of Fit	620.36	11	56.40	0.6347	0.7542	not significant
Pure Error	444.30	5	88.86			
Cor Total	1787.89	19				

The analysis of variance of the response, MRR is summarized in Table 2 and the significant 2 Factorial model is shown. The feed is the dominant factor affecting MRR. The Model F-value of 1.19 suggests the model is not significant relative to the noise. There is a 37.16% chance that huge F-value could arise due to noise. P-values <0.0500 signpost model terms are significant. In this item there are no significant model terms. Values >0.1000 specify the model terms are not significant. The Lack of Fit F-value of 1.67 denotes the Lack of Fit is not significant relative to the pure error. There is a 29.78% chance that a Lack of Fit F-value will occur. Non-significant lack of fit is good.

Table 2. ANOVA Table for Material Removal Rate.

Source	Sum of Squares	df	Mean Square	F-value	p-value	
Model	21220.99	6	3536.83	1.19	0.3716	not significant
A-Cutting Speed	925.89	1	925.89	0.3106	0.5868	
B-Feed	4307.12	1	4307.12	1.44	0.2508	
C-Depth of cut	8.35	1	8.35	0.0028	0.9586	
AB	11615.17	1	11615.17	3.90	0.0700	
AC	4352.91	1	4352.91	1.46	0.2484	
BC	11.54	1	11.54	0.0039	0.9513	
Residual	38750.36	13	2980.80			
Lack of Fit	28174.54	8	3521.82	1.67	0.2978	not significant
Pure Error	10575.82	5	2115.16			
Cor Total	59971.35	19				

2.4. Model validation

Linear model Residual and Leverage values for Surface Roughness are shown in Table 3. It clearly depicts the actual and predicted values. The aim of this step is to predict and verify the improvement of the response using the optimal levels of the turning process parameters. Adequate Precision measures the signal to noise ratio. A ratio larger than 4 is necessary. The experimental ratio of 6.225 specifies an acceptable signal. This model can be used to circumnavigate the design space.

Table 3. Residual and Leverage values for Surface Roughness.

Run Order	Actual Value	Predicted Value	Residual	Leverage	Internally Studentized Residuals	Externally Studentized Residuals	Cook's Distance	Influence on Fitted Value DFFITS
1	32.00	16.36	15.64	0.050	1.967	2.187	0.051	0.502
2	3.10	17.03	-13.93	0.257	-1.981	-2.207	0.339	-1.299
3	13.50	16.36	-2.86	0.050	-0.360	-0.350	0.002	-0.080

4	12.00	11.84	0.1584	0.257	0.023	0.022	0.000	0.013
5	4.73	11.90	-7.17	0.270	-1.029	-1.031	0.098	-0.626
6	3.54	16.36	-12.82	0.050	-1.613	-1.706	0.034	-0.391
7	30.12	25.41	4.71	0.270	0.676	0.664	0.042	0.404
8	30.75	20.82	9.93	0.270	1.424	1.476	0.187	0.897
9	19.56	20.88	-1.32	0.257	-0.188	-0.182	0.003	-0.107
10	16.30	27.72	-11.42	0.257	-1.624	-1.720	0.228	-1.012
11	15.00	15.70	-0.6975	0.257	-0.099	-0.096	0.001	-0.057
12	32.78	26.20	6.58	0.270	0.945	0.941	0.082	0.572
13	11.56	5.01	6.55	0.257	0.932	0.928	0.075	0.546
14	20.30	16.36	3.94	0.050	0.495	0.483	0.003	0.111
15	4.81	7.32	-2.51	0.270	-0.360	-0.350	0.012	-0.212
16	3.98	6.53	-2.55	0.270	-0.365	-0.355	0.012	-0.216
17	15.40	12.69	2.71	0.270	0.388	0.377	0.014	0.229
18	16.80	16.36	0.4384	0.050	0.055	0.053	0.000	0.012
19	21.50	16.36	5.14	0.050	0.646	0.634	0.005	0.145
20	19.50	20.03	-0.5303	0.270	-0.076	-0.074	0.001	-0.045

Two factorial model Residual and Leverage values for MRR are shown in Table 4. It clearly depicts the actual and predicted values in relation with Cook's distance.

Table 4. Residual and Leverage values for MRR.

Run Order	Actual Value	Predicted Value	Residual	Leverage	Internally Studentized Residuals	Externally Studentized Residuals	Cook's Distance	Influence on Fitted Value DFFITS
1	126.00	177.92	-51.92	0.050	-0.976	-0.974	0.007	-0.223
2	257.00	191.77	65.23	0.257	1.386	1.443	0.095	0.849
3	160.00	177.92	-17.92	0.050	-0.337	-0.325	0.001	-0.075
4	155.00	179.23	-24.23	0.257	-0.515	-0.500	0.013	-0.294
5	136.80	167.12	-30.32	0.645	-0.932	-0.927	0.225	-1.248
6	148.40	177.92	-29.52	0.050	-0.555	-0.539	0.002	-0.124
7	190.00	124.03	65.97	0.645	2.027	2.355	1.065(1)	3.172(1)
8	250.20	220.67	29.53	0.645	0.907	0.901	0.213	1.213
9	149.00	176.60	-27.60	0.257	-0.587	-0.571	0.017	-0.336
10	123.00	207.79	-84.79	0.257	-1.802	-1.998	0.160	-1.176
11	146.00	164.07	-18.07	0.257	-0.384	-0.371	0.007	-0.218
12	255.50	263.36	-7.86	0.645	-0.241	-0.233	0.015	-0.313
13	200.00	148.05	51.95	0.257	1.104	1.114	0.060	0.655
14	210.00	177.92	32.08	0.050	0.603	0.587	0.003	0.135
15	39.78	106.54	-66.76	0.645	-2.051	-2.397	1.091(1)	-3.228(1)
16	220.00	212.94	7.06	0.645	0.217	0.209	0.012	0.281
17	127.49	154.04	-26.55	0.645	-0.816	-0.805	0.172	-1.084
18	224.60	177.92	46.68	0.050	0.877	0.869	0.006	0.199

19	239.20	177.92	61.28	0.050	1.152	1.168	0.010	0.268
20	200.40	174.65	25.75	0.645	0.791	0.779	0.162	1.050

2.5. *Results*

The normal probability plots of the residuals and the plots of the residuals versus the predicted response for surface roughness and MRR are shown in Figure 1, 2, 5 and 6 respectively. The Figure 1, 5 revealed that the residuals fall on a straight line indicating that the errors are disseminated ordinarily. The Figure 2, 6 revealed that they have no obvious pattern and unusual structure. This implies that the models proposed are adequate and there is no reason to suspect any violation of the independence or constant variance assumption. The interaction effects between the three factors are shown in the Figure 3. From the figures, it can be inferred that the three factors speed, feed and depth of cut are relatively independent of each other, i.e. variation of one factor doesn't significantly affect the effects of other factors on the response.

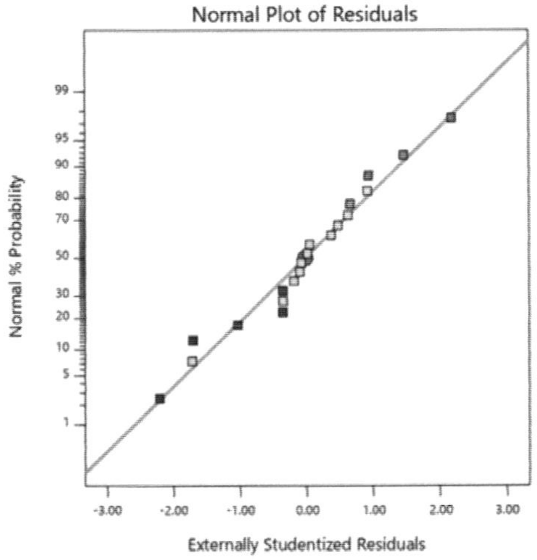

Figure 1. Normal probability plot of residuals for Ra data.

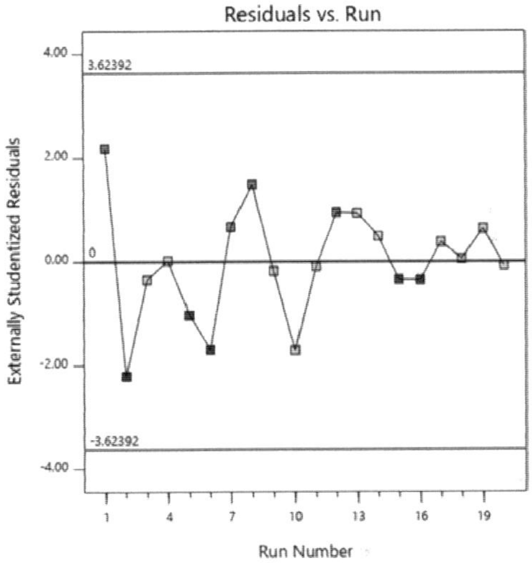

Figure 2. Residuals Vs Run for Ra data.

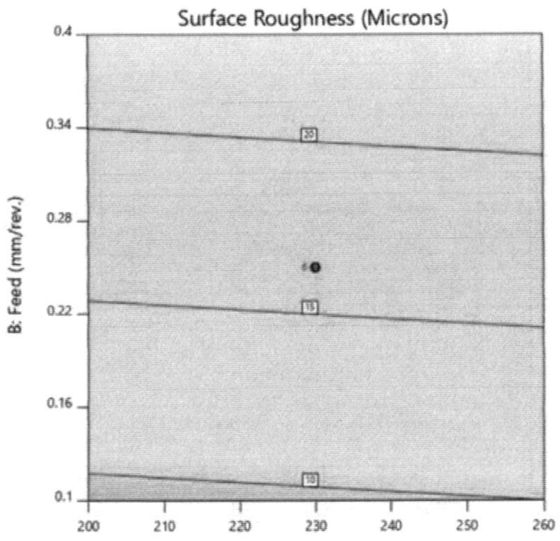

Figure 3. Contour plot for Ra data.

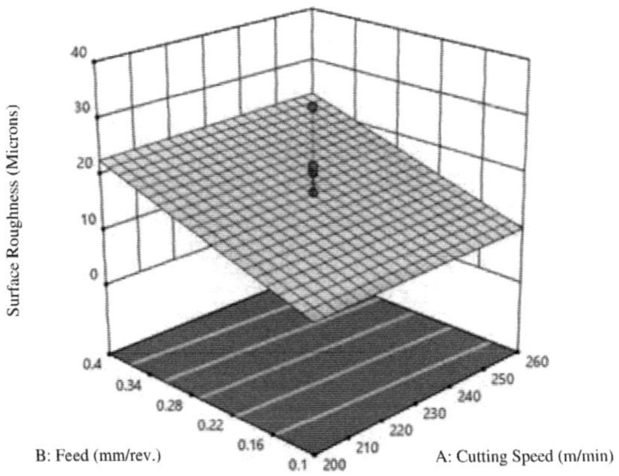

Figure 4. 3D surface graph for Ra data.

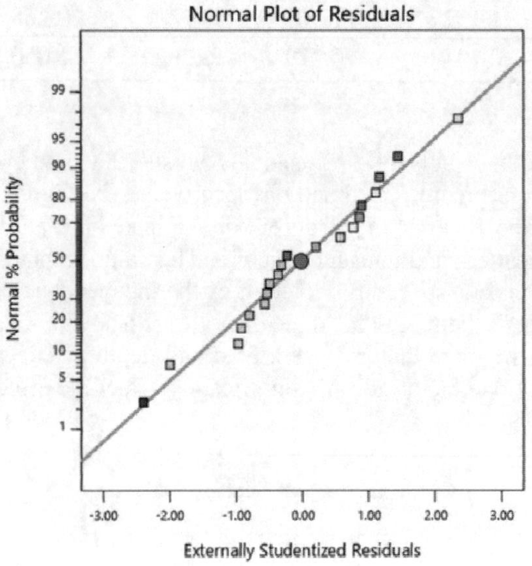

Figure 5. Normal probability plot of residuals for MRR data.

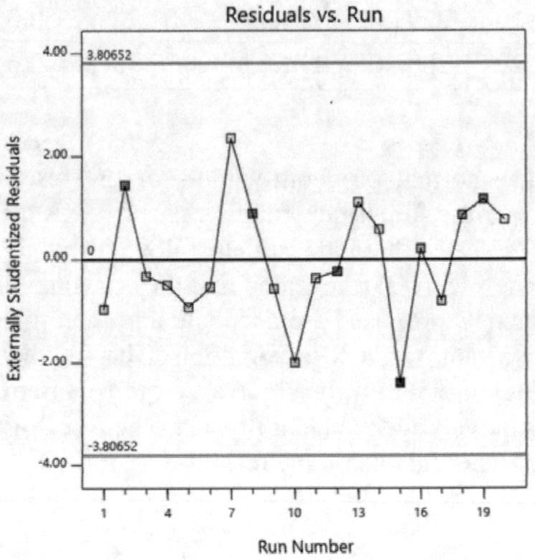

Figure 6. Residuals Vs Run for MRR data.

A 2D contour plot was generated for the responses at any two independent variables while keeping the others at their 0 level. The contour for the response surface for surface roughness and MRR is shown in Figure 3 and 7. It is clear from the figure that at medium feed rate, the best surface finish is obtainable and when the cutting speed is somewhere at middle range experimented. It is also clear that the surface roughness rises with growing feed. The observation made for MRR is also an increasing trend. MRR escalates with the upsurge in cutting peed. It is clearly visible from Figure 7 as red portion. The MRR is dependent on the cutting speed which is a significant factor.

The RSM was used to study the three-dimensional response plots, which were generated from the effects of the three variables on surface roughness and material removal rate. The response values are now investigated, by building contour and 3D surface plots, to study the effects of the different factors on the response. Response optimization is performed using the derived regression model, and the optimal values are determined. Figure 4 and 8 shows the 3D graph of the effect of cutting speed and feed on the surface roughness and MRR. Both have curvilinear profile in accordance to the model fitted.

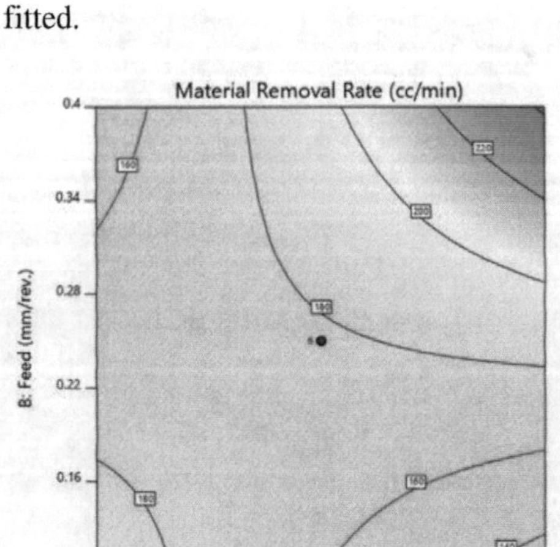

Figure 7. Contour plot for MRR data

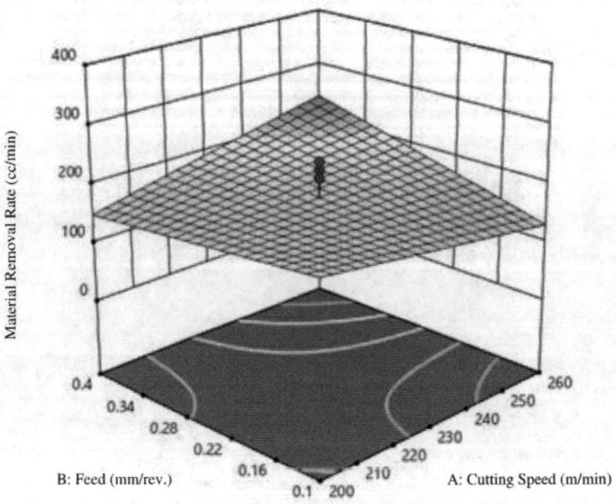

Figure 8. 3D surface graph for MRR.

From the obtained regression model, the optimum values are determined. If the objective is to minimize surface roughness (Ra) only, MRR is not optimized. From the Figure 4, it can be inferred that cutting speed and feed have a linear relationship with surface roughness (Ra). When the objective is to maximize the material removal rate (MRR) only, Ra is not optimized. The optimization plot for the above condition is provided in the Figure 8. Material removal rate is linearly dependent on the three factors-speed, feed and depth of cut, which is verified from the graph. Also from the plot, Maximum MRR is obtained when all three cutting parameters are at maximum value.

The goal is to minimize surface roughness (Ra) and maximize material removal rate (MRR), Clearly, the optimum response is obtained at V = 260 m/min, f = 0.1935 mm/ rev, d = 2.5 mm. The consequences of the experimental data are used for predicting the outcome of assorted input machining parameters such as cutting speed, feed and depth of cut on the surface roughness when machining POM. Cutting speed has the strongest effect on the surface roughness among the selected parameters; it is inversely proportional to the response. It is found that the roughness could be controlled in the design stage which is the most effective and inexpensive way.

2.6. *Confirmation tests*

In order to validate the adequacy of the model developed, confirmation run experiments are performed (Table 5). The predicted values and the associated prediction interval are centred on the model developed previously. The predicted model mean value and the actual experimental values are compared. Two sided with confidence level 95% and population 99% are considered. All these values are presented in Table 5.

Table 5. Confirmation test point prediction

Response	Predicted Mean	Predicted Median	Observed	Std. Dev.	SE Mean	95% CI low for Mean	95% CI high for Mean	95% TI low for 99% Pop	95% TI high for 99% Pop
Surface Roughness	17.0257	17.0257	17.031	8.15728	4.1362	8.25733	25.794	-19.9822	54.0336
Material Removal Rate	191.766	191.766	191.61	54.5967	27.6836	131.959	251.573	-66.1549	449.687

3. Conclusion

In this paper a methodology based on RSM optimization, for identifying the optimal parameters for machining a work piece are presented. The advantageous properties of POM and its applications are studied. The experimental study is conducted and the effects of different factors named speed, feed and depth of cut, on surface roughness are reported. Response optimization is performed, to obtain minimum surface roughness and maximum material removal rate. The optimization results obtained in this work confirm that the proposed optimization method is a very useful tool for multi-objective optimization of machining parameters. The model generated by RSM design for surface roughness and material removal rate has been found suitable to predict the performance of the POM material. The deviation under the confidence level of 95% in the data values of experimentally collected and that generated from the RSM linear model assure the accuracy of the RSM model.

4. References

[1] Juan C Campos Rubio, Tulio H Panzera and Fabrizio Scarpa 2014 Machining behaviour of three high-performance engineering plastics *Proc., of the Inst. of Mech. Engrs, Part B: J. of Engg. Manufacture* 1–10

[2] Endo H and Marui E 2006 Small-hole drilling in engineering plastics sheet and its accuracy estimation *Int. J. Mach. Tool Manufacture* **46(6)** 575–579

[3] Fetecau C, Stan F, Munteanu A and Popal V 2008 Machining and surface integrity of polymeric materials *Int. J. Mater. Form* **1(1)** 515–518

[4] Davim JP and Mata FJ 2007 A Comparative evaluation of the turning of reinforced and unreinforced polyamide *Int. J. Adv. Manuf. Technology* **33 (9–10)** 911–914

[5] Campos Rubio JC, Panzera TH, Abrao AM, Paulo E. Faria and Paulo Davim 2011 Effects of high speed in the drilling of glass whisker reinforced polyamide composites (PA66 GF30): statistical analysis of the roughness parameters *J. Compos. Materials* **45(13)** 1395–1402

[6] Sreejith PS, Krishnamurthy R, Malhotra SK, Narayanasamy K and Malhotra SK 2000 Evaluation of PCD tool performance during machining of carbon/phenolic ablative composites *J. Mater. Prod. Technology* **104(1)** 53–58

[7] Davim JP and Reis P 2004 Machinability study on composite (polyether-ether-ketone reinforced with 30% glass fibre– PEEK GF 30) using polycrystalline diamond (PCD) and cemented carbide (K20) tools *Int. J. Adv. Manuf. Technology* **23(5–6)** 412–418

[8] Hocheng H and Tsao CC 2006 Effects of special drill bits on drilling-induced delamination of composite materials *Int. J. Mach. Tool Manufacture* **46(12–13)** 1403–1416

[9] Endo H and Marui E 2005 Effect of the specimen geometry on wear – combination of polyacetal (POM) and carbon steel for machine structures *Wear* **258(10)** 1525–1530

[10] Akira Kobayashi and Kenji Hirakawa 2006 Ultraprecision Machining of Plastics Part 1 Polymethyl Methacrylate *Polymer-Plastic Tech. Engineering* **22(1)1984** 15-25

[11] Jagtap K and Pawade R 2014 Experimental investigation on the influence of cutting parameters on surface quality obtained in SPDT of PMMA *Intr. J. of Adv. Design and Manuf. Technology* **7(2)** 53- 58

[12] Mehdi Moghri, Milos Madic, Mostafa Omidi and Masoud Farahnakian 2014 Surface roughness optimization of polyamide-6/Nanoclay Nanocomposities using artifical neural network: genetic algorithm approach *The Scientific World Journal* **2014** 1-7

[13] Paizun AA, Fathullah M, Abdullah MMA, Shayfull Z and Faheem Tahir 2019 Surface roughness optimization on rubberized fly ash geopolymer in lathe operation using Taguchi method *AIP Conference Proceedings* vol **2129** p 02018

[14] Manohar M, Jomy Joseph, Selvaraj T and Sivakumar D 2013 Application of Box Behnken Design to optimize the parameters for turning Inconel 718 using coated carbide tools *Int. J. of Scient. & Engg. Research* **4(4)** 620

[15] Arifin N, Yusoff H, Sudin I, Kadir AZA, Ali R, Yacob S, Arshad A and Ismail SA 2016 Study the effect of CNC milling parameters on surface roughness of POM material *ARPN J. Engg. and Applied Sciences* **11(10)** 6611-6614

[16] Montgomery D C 1997 *Design and Analysis of Experiment* (New York : Wiley)

Vibration Response Simulation on a High Risk Building with Laminated Rubber Bearing Base Isolation Subjected to Earthquake Excitation

F R Hardian[1] **and W Nirbito**[2, *]

Department of Mechanical Engineering, Faculty of Engineering, Universitas Indonesia, Kampus Baru UI Depok, Depok 16424, Indonesia

Email: [1] firhanhardian@gmail.com; [2] bito@eng.ui.ac.id (* Corresponding Author)

Abstract. In a high risk building like hospitals, laboratories, or school, the response of the building due to earthquake might be very dangerous. To reduce victims and damages in infrastructure, a type of base isolation is installed to the foundation of the building. The base isolation, in form of a laminated rubber bearing, produces a horizontal movement of the structure when subjected to earthquake forces in order to keep the people and equipment safe. Some calculations of the response vibration parameters are observed to determine the optimal characteristics of the base isolation. The Keq Value is 415,120,000 N/m and the total mass of the building is 739,534,249.7 kg as well as the highest vibration force is 4,015,079,348.47 N. The aim of this research is to simulate the lateral motion of the building with base isolation, due to earthquake forces and decide whether the response is as expected of the design.

1. Introduction

Indonesia is located in between three continental plates, as well as in the center of the ring of fire are where most of the planet's active volcanoes are also located. With a long history of massive earthquakes and the risk of possible future earthquakes, a preventive action is used in the form of base isolations. Base isolation is an earthquake damage reduction method that is applied on the foundation of a building. The main function of base isolation is to reduce the response vibration excited by the earthquake to minimize victims and damages. One of the popular types of base isolation is the laminated rubber bearing, which consists of layers of steel and rubber plates. The laminated rubber bearings absorb the earthquake force without continuing it to the rest of the structure above [1], thus creating a lateral movement instead of a cantilever movement. The aim of this study is to simulate the response of a high risk building equipped with laminated rubber bearing base isolation o find out whether the base isolation can work to its designated function, as well as to compare the value of displacement based on two types of system simplification.

2. Analysis model

2.1. Specifications of base isolation

The actual laminated rubber bearings used in the building studied has the following specifications presented in Table 1. The data will be used for simplification on the simulation stage. There are 161 bearings divided in four types of different diameter sizes distributed on the foundation of the building. The types of rubber used are the high damping rubber bearing which has no lead core in the center.

Table 1. Design characteristics of base isolation

Component		HH800	HH1100	HH1200	HH1300
Outer Diameter	d_0 [mm]	800	1100	1200	1300
Shear Modulus	Geq [N/mm^{2}]	0.620	0.620	0.620	0.620
Thickness of rubber layer	Tr [mm]	5.4	7.4	8.0	8.7
Number of rubber layer	N	37	27	25	23
Height	H [mm]	422.2	390.2	385.6	376.9
Spring Constant	10^3 x [kN/m]	1.56	2.94	3.50	4.11

The base isolation is modeled as in Figure 1. With the green layer being the rubbers and the black layers being the steel plates.

Figure 1. Base isolation modeled in autodesk inventor.

3. Simulation parameters

To get higher accuracy in the simulation, some relevant data are needed. The building mass (m) calculated from ANSYS mass calculation feature obtained the value of 739,534,249.7 kg. In addition to the mass, the earthquake excitation force value is also needed. The value of the force is obtained from the earthquake response spectrum graph shown in Figure 2 from which the building is located using the following Eq. 1.

$$F = m \cdot a \tag{1}$$

The maximum acceleration value is chosen to get maximum response value, which is 0.554 times gravitational acceleration. Thus the value of excitation force is 4,015,079,348.47 N. Other parameter needed is the frequency of the earthquake and that is 1.75 Hz also taken from the earthquake response spectrum, due to the fact that it is the lowest frequency in the highest acceleration, therefore creating the highest response.

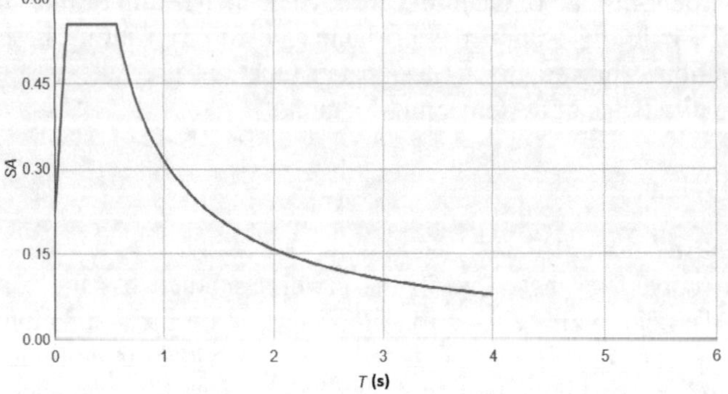

Figure 2. Earthquake response spectrum in which the building studied is located.

IOP Conf. Series: Materials Science and Engineering **893** (2020) 012006 doi:10.1088/1757-899X/893/1/012006

4. Simulation preparations

In this study, two types of simulation are ran. The first is the simple simulation that simplified the 161 base isolations in the building into just one with mean values for all the dimensions. The second simulation is the unified simulation that sum up the total diameter and created one base isolation in a larger scale using comparative equation. The simulations are done using ANSYS Mechanical with the Keq value of 415,120,000 N/m. The simulation type is using Harmonic Response function with the calculated parameters inputted [2]. The details of the dimensions are listed in Table 2.

Table 2. The specifications for the model in two simulations

Specifications	Simple	Unified
Outer Diameter [mm]	1.01	13.02
Inner Diameter [mm]	40.21	510
Thickness of one steel plate [mm]	6.81	87
Thickness of one rubber layer [mm]	6.79	82
Number of steel plate	30	30
Number of rubber layer	31	31
Total height [mm]	401.4	5160

For a clearer visualization, a structural steel base [3] is added at the top and bottom of the base isolation model with the dimension of 20,000 x 20,000 x 2,000 mm. The top base is assumed as the building and the bottom base as the ground, the dimensions are the same so the visualization can be focused on the rubber part of the model. A distributed mass of 739,534,249.7 kg is applied vertically to the top base and an excitation force of 4,015,079,348.47 N is applied horizontally at the bottom base. A fixed support was applied at the bottom of the model [4] and meshing was done [5]. The pre-simulated model is shown in Figure 3.

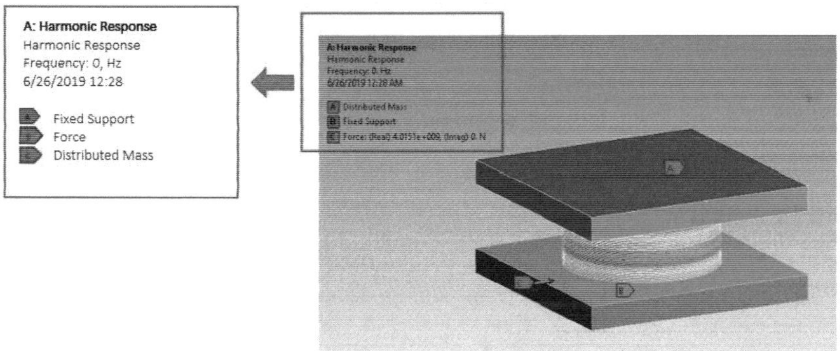

Figure 3. The pre simulated model for the simulations.

5. Result and discussion

The results of the simulations are the vertical and horizontal displacement value as well as the visualization of the motion of the building subjected to earthquake excitation. The results shown are from the model created using specifications in Table 2. The visualizations are shown in Figure 4 and Figure 5. From both simulations, the results show that the displacement in the horizontal direction is greater than the horizontal displacement up to three times. The exact value of the displacements are shown in Table 3.

IOP Conf. Series: Materials Science and Engineering **893** (2020) 012006 doi:10.1088/1757-899X/893/1/012006

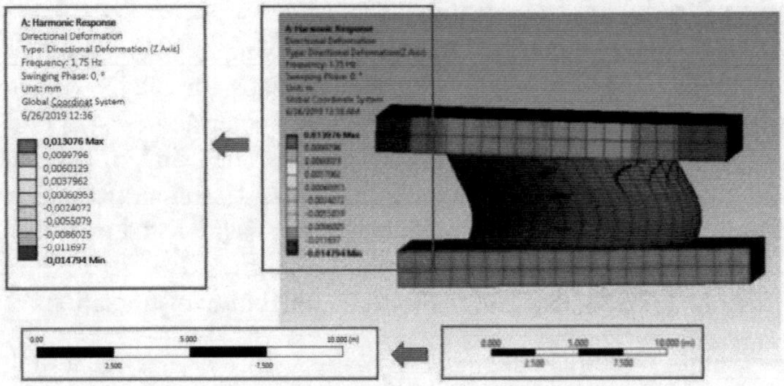

Figure 4. The visuals of displacement on the simple.

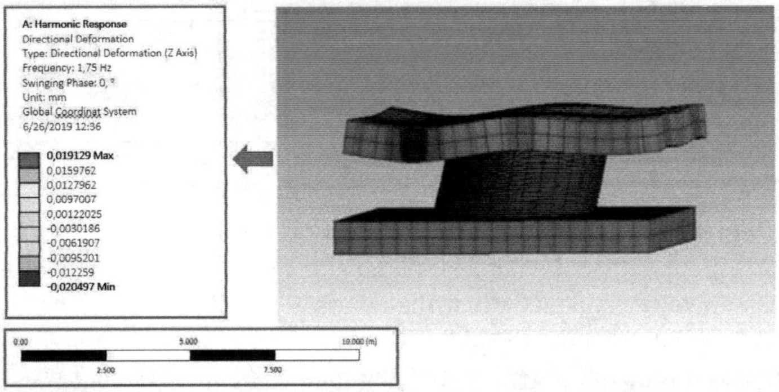

Figure 5. The visuals of displacement on the unified simulation.

Table 3. Value of displacements from the simulations

Simulation	Vertical Displacement [mm]	Horizontal Displacement [mm]
Simple	14	47
Unified	19	60

To compare, the unified simulation creates a wavy surface on the top base that acts as the building, it can be seen that the value of displacement has very small difference, unlike the simple simulation that keeps the building in shape. This happens because the dimension of the unified simulation model is far bigger than the simple simulation, thus creating a smaller ratio of size to force. Therefore the base isolations resisted the force better at the bottom to middle part, creating a wavy pattern at the top [6].

A mechanical engineering approach is used to obtain the frequency ratio value using the magnification ratio vs frequency ratio graph shown in Figure 6. The natural frequency value of system is obtained using the modal analysis function in ANSYS and that it 0.112 Hz (f_n), and 1.75 Hz excitation frequency (f) value from earthquake response spectrum. Therefore the value of frequency ratio, calculated using Eq. 2 is 15.62 Hz. That value shows that both of the simulation results have very low magnification factor, creating a safe system for the people and equipment inside, also stating that the base isolations are functioning to its design [7].

$$\beta = \frac{f}{f_n} \quad (2)$$

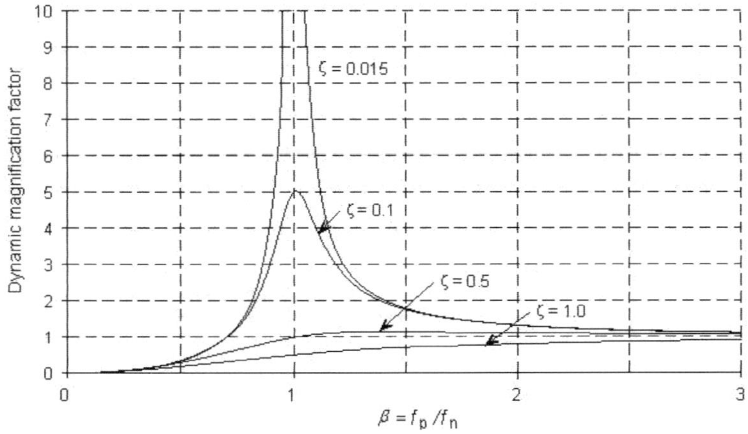

Figure 6. Magnification factor vs frequency ratio

6. Summary

Base isolation creates response that is hand in hand with the designed function, when an earthquake occur the response will be in a lateral horizontal motion. With a mass of 739,534,249.7 kg and excitation force of 4,015,079,348.47 N The simulations done have shown that the dominant response of the building is in the horizontal direction, with the value from the simple simulation and unified simulation are 47 and 60 mm respectively. The ratio of the horizontal and vertical displacements are also the same which is 1:3. The building in this study with all the base isolation system is considered safe due to the fact that the magnification factor of the system is very low. From this study, it can be concluded that the simplifications used are showing pretty similar results and can be used for future preventive simulation on base isolation.

7. References

[1] Febymol K. B., Rohini G. Nair, Finite Element Analysis of Elastomeric Bearing: submitted to International Research Journal of Advanced Engineering and Science, Volume 2, Issue 2. (2017)

[2] Daniela O. F, Mihai Budescu, Sergiu Baetu, Vlad Fediuc, Finite Element Modelling of Elastomeric Bearings, "Gheorghe Asachi" Technical University of Iaşi Faculty of Civil Engineering and Building Services. (2015)

[3] Manasa M. S., Alice Mathai, Performance of Lead Rubber Bearing as a Base Isolator: submitted to International Journal of Science Technology & Engineering, Volume 3, Issue 11. (2017)

[4] Makoto Ohsaki et al. Finite-Element Analysis of Laminated Rubber Bearing of Building Frame under Seismic Excitation, Department of Architecture, Hiroshima University. (2015)

[5] Animesh Das, Anjan Dutta, S.K. Deb, Modeling of Fiber-Reinforced Elastomeric Base Isolators: submitted to 15th World Conference on Earthquake Engineering, Lisbon. (2012)

[6] Kelly, J. M., Calabrese, A., & Serino, G. , Design Criteria For Fiber Reinforced Rubber Bearings: submitted to 15th World Conference on Earthquake Engineering, Lisbon. (2012)

[7] Assainar, N.K., Sreemahadevan, Nonlinear Analysis of High Damping Rubber Bearing with Different Material & Geometry: submitted to International Journal of scientific & Engineering Research (2016)

Acknowledgments

This research is supported and funded by DRPM UI with grand PITTA No. DLG-12/UN2.R3/PPM.00.00/2019. We thank Universitas Indonesia Teaching Hospital and Henki Ashadi, PhD who provided insights that greatly assisted the study, although they may not agree with all of the interpretations or conclusions of this paper.

Optimization of Laminate Topology for Multilayered Composite Structures Subjected To Buckling and Failure Constraints

M Augustyn[1, *] and A Muc[1]

[1] Institute of Machine Design, Faculty of Mechanical Engineering, Cracow University of Technology, Kraków, Poland.

E-mail: augustyn@mech.pk.edu.pl

Abstract. In the present paper the combinatorial optimization problems dealing with laminate stacking sequence optimization are formulated and studied. Two types of design variables are introduced: 1) classical, discrete and 2) new, continuous ones. In order to obtain the optimal solution a classical genetic algorithm is applied. We are looking for the maximal buckling loads for 2-D multilayered structures subjected to bimodality constraints and FPF constraints. Two special buckling problems are analyzed in details, i.e. buckling of compressed rectangular plates and of cylindrical shells under an external pressure. The results demonstrate the effectiveness and advantages of the use of continuous design variables.

1. Introduction

Material properties of composite materials strongly depend on fibre orientations in each layer, so that the optimization of laminate topology becomes one of the most important problems. On the other hand, for isotropic structures a large number of failure modes exist comparing with those for isotropic constructions. For instance, one can distinguish the global failure mode i.e. buckling and the local modes in the form of *first-ply-failure*, *delamination with local buckling* and *matrix* or *fibre breakage*. Nowadays, it is possible to find the optimal topology of the laminate independently for each failure form mentioned above. The problem of searching for the optimal laminate topology, when several failure modes are taken into account is still unsolved in a general manner. The current work is devoted to this question. We will present with details different mathematical formulation of the optimization problem. This problem will be stated as single-modal and multi-modal with various constraints. The optimization will be formulated for buckling of plates and composite shells with different failure forms (*FPF and delamination*), that occurred at the same time. The influence of modal and multi-modal buckling forms on optimal laminate configuration will be also presented.

The aim of the present work is two-fold:

- To discuss various formulations of different optimization problems with the aid of new, continuous type of design variables introduced instead of classical fibre orientations and stacking sequences describing mechanical properties of multilayered composite laminates,
- To obtain the optimal solution using Borland BILDER C^{++} and genetic algorithms provided by the commercial *Evolver* package in order to demonstrate the possibility of solving completely different problems in the identical way.

We shall not dwell herein on the repetition of the existing literature in this area. In our opinion it is worth to mention only Refs. [1, 2] showing the optimization methods used in the problems of composite laminates as well as the general ideas of genetic algorithms.

For isotropic and multilayered composite structures different problems dealing with stacking sequence optimization of 2D (beams, plates, shells) problems are formulated, discussed in details and solved in Refs [3-7]. A special attention is focused on the reduction of total number of design variables characterizing optimization objectives.

2. Definition of design variables

In 2-D approach a laminate configuration is expressed uniquely by the appropriate terms of the A, B and D matrices. Topological variables defining the connectivity of particular structural elements in the structure; in the paper it denotes the stacking sequence of the individual layers in the laminate, are understood in the sense of the sequence of layers having prescribed discrete fibre orientations θ_i in each individual ply; commonly, it is assumed that the thickness of individual plies are identical, i.e. ti = t/N, where t denotes the total thickness of the laminate, whereas N means the total number of plies, and i = 1,2,...,N. In order to assure a great flexibility and generality in formulation of various optimization problems different types of the above-mentioned discrete design variables must be represented in a similar unified manner, i.e. each design variable must be coded as a finite string of digits in the classical GA. Let us note that the angle-ply anti-symmetric laminates are considered only, however, it can be easily extended for arbitrary laminate configuration. Using the classical method of coding 1 represents $0°_2$, 2 – $\pm 45°_2$, 3 – $90°_2$. Each design variable s representing fibre orientation (i.e. 1, 2, and 3) is coded as binary number and called as a gene. The sequence {1,2,1,3} is called as a chromosome. Such a representation is not very convenient for optimization problems since there are a lot of design variables (increasing with the total number of plies N) and in addition various stacking sequences are described by the identical values of the A, B, D matrices. Therefore, we propose to adopt herein a special type of continuous variables $x_r^{\{A,D\}}$ introduced by Muc [3, 4] that are completely different than those used by Miki [8]. The new design variables represent triangles in the design space, however there is no unique mapping between spaces x_r^A and x_r^D. For the assumed laminate configuration the B matrix is identically equal to zero whereas the terms of the A and D matrices can be written in the following way:

$$A_{11}=t(U_1-U_3)+\frac{4t}{N}U_2(x_1^A-x_3^A)+\frac{8t}{N}U_3(x_1^A+x_3^A),$$

$$D_{11}=\frac{t^3}{12}(U_1-U_3)+\frac{1}{12}\left(\frac{4t}{N}\right)^3 U_2(x_1^A-x_3^A)+\frac{1}{6}\left(\frac{4t}{N}\right)^3 U_3(x_1^A+x_3^A), \tag{1}$$

$$x_r^{\{A,D\}}=\sum_{k=1}^{N/4}\left\{1,\frac{3k(k-1)+1}{3}\right\}\cos 2\theta\,\Xi(\alpha_r),\quad \Xi(\alpha_r)=\begin{cases}1 & \text{gdy} \quad \alpha_r=\theta \\ 0 & \text{gdy} \quad \alpha_r\neq\theta\end{cases},\quad \alpha_r=90^0\frac{r-1}{3},\quad r=1,2,3$$

3. Formulation of the optimization problem

The optimization problem is formulated as follows:

$$\underset{S}{Max}\left(\underset{m,n}{Min}\,\lambda_b\right), \tag{2}$$

where λ_b is a critical multiplier of loading (understood in the sense of a global buckling), *S* denotes the vector of design variables representing fibre orientations in the laminate in a discrete or continuous manner given by relations (1), and *m*, *n* are numbers of half-waves in two perpendicular directions corresponding to a plate/shell co-ordinate system. The analysed problem may be subjected to various subsidiary constraints written in the following form:

Bimodal constraints:
$$\lambda_b[m,n]\leq\lambda_b[m+1(m),n(n+1)], \tag{3}$$

Failure constraints: $$\lambda_{failure} \le \lambda_b \tag{4}$$

The constraint (3) presents two conditions for each wave number in buckling independently, whereas the relations (4) demonstrates the possibility of introduction of various failure modes in the form of the matrix cracking, fibre breakage, delaminations etc. The definition of the $\lambda_{failure}$ multiplier is directly connected with the analyses failure mode.

To illustrate the effectiveness of the proposed method of solution two buckling problems will be solved in details. The first one deals with the buckling problem of bi-axially compressed, rectangular plates, where (Ref. [9]):

$$\lambda_b = \frac{(m\pi / a)^2}{P_x\left(1 + k\beta_m^2\right)}\left[D_{11} + 2(D_{12} + 2D_{66})\beta_m^2 + D_{22}\beta_m^4\right], \quad \beta_m = (na)/(mb), \qquad k = P_y / P_x, \tag{5}$$

and in the second buckling of cylindrical shells under external pressure is considered (Ref. [10]):

$$\lambda_b p\frac{n^2}{R} = K_{33} + \frac{2K_{12}K_{13}K_{23} - K_{11}K_{23}^2 - K_{22}K_{13}^2}{K_{11}K_{22} - K_{12}^2},$$

$$K_{11} = A_{11}\left(\frac{m\pi}{L}\right)^2 + A_{66}\left(\frac{n}{R}\right)^2, \quad K_{12} = (A_{12} + A_{66})\left(\frac{m\pi}{L}\right)\left(\frac{n}{R}\right), \qquad K_{22} = A_{22}\left(\frac{n}{R}\right)^2 + A_{66}\left(\frac{m\pi}{L}\right)^2,$$

$$K_{13} = \frac{A_{12}}{R}\left(\frac{m\pi}{L}\right), \; K_{23} = \frac{A_{22}}{R}\left(\frac{n}{R}\right), \tag{6}$$

$$K_{33} = D_{11}\left(\frac{m\pi}{L}\right)^4 + 2\left(D_{12} + 2D_{66}\right)\left(\frac{m\pi}{L}\right)^2\left(\frac{n}{R}\right)^2 + D_{22}\left(\frac{n}{R}\right)^4 + \frac{A_{22}}{R^2}$$

where p denotes an external, uniform pressure, and R and L are the shell radius and length, respectively.

4. Numerical solutions

To find the optimal solutions of the problem given by equations (2)-(4) in the most general manner the package Evolver have been employed. It is conjugated with the C^{++} Builder to evaluate different forms of the buckling load parameter as well as constraint conditions. The analysis has been started from the discrete definition of the design variables that are assumed to take the following form: $0°_2$ or $\pm 45°_2$ or $90°_2$. The laminate of the thickness equal to 6.096 mm is made of 48 individual plies (but according to the assumed laminate symmetry 12 of them only constitute the vector of design variables) having the following material properties: E_1=127.6 GPa, E_2=13.03 GPa, G_{12}=6.41 GPa, ν_{12}=0.3. In all cases considered the population have been identical and possessed 70 individuals; the one-point crossover probability is equal to 0.5, whereas the mutation parameter is equal to 0.01. To compare the optimal solutions the initial population has been identical and made of genes having orientations $0°_2$ only. The results of 100 independent searching for the optimal stacking sequence of rectangular plates are presented in Figure 1. As it may be seen a scatter of optimal values of buckling load factor is observed and in addition the maximal value obtained with the use of continuous variables (5) does not correspond exactly to those evaluated in the numerical way (the classical GA).

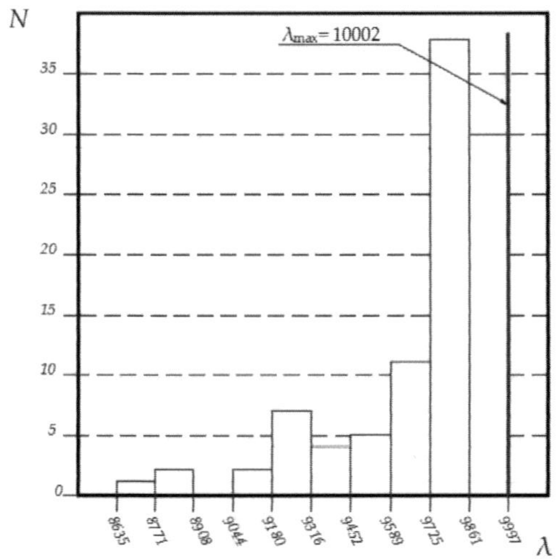

Figure 1. The bar chart plot of optimal results for rectangular plates (a/b=3, n=1, m=3) subjected to bimodal and first-ply-failure constraints.

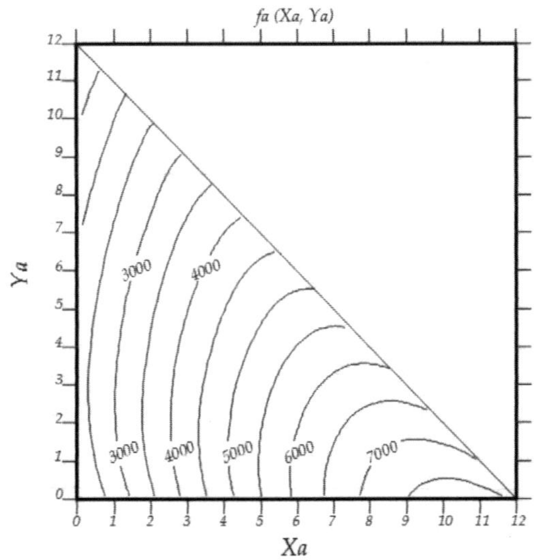

Figure 2. The contour plot of the membrane part of the buckling load factor – a cylindrical shell (L/D=5/3, n=6, m=1) under external pressure.

This is a typical result which can be noticed using classical GA for multilayered structures. The terms of the D matrix are responsible for buckling results, and the A matrix for the plate first-ply-failure. Therefore it is impossible to obtain always identical optimal stacking sequences, mainly due to the lack of the unique mapping between the A and D matrices terms.

It is better demonstrated in the contour plots of the buckling load factor for cylindrical shells under external uniform pressure – see Figure 2 and 3. Figure 2 presents the distributions of the part of the buckling load factor corresponding to the membrane terms in the relation (6), whereas Figure 3 is a plot of the latter bending part. Even looking for the optimum given by equation (2) with no constraints it is impossible to find directly the optimal laminate configuration.

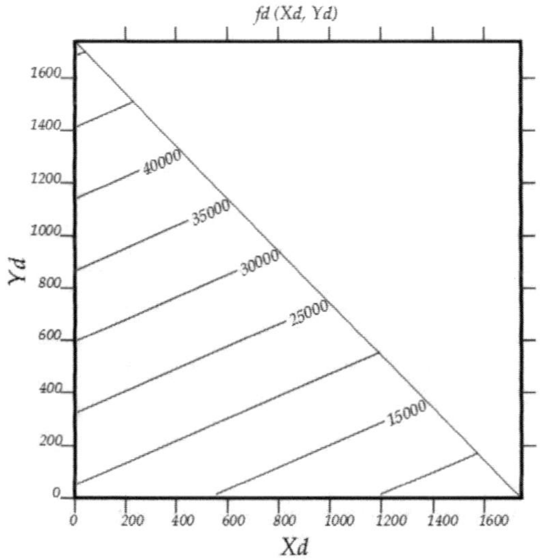

Figure 3. The contour plot of the bending part of the buckling load factor – a cylindrical shell (L/D=5/3, n=6, m=1) under external pressure.

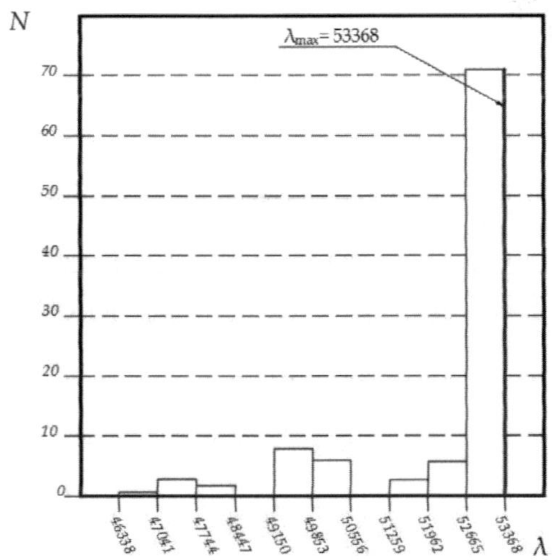

Figure 4. The bar chart plot of optimal results for cylindrical shells.

IOP Conf. Series: Materials Science and Engineering **893** (2020) 012007 doi:10.1088/1757-899X/893/1/012007

However, the optimum occurs at the boundary of the design variable domain. Similarly as previously for multilayered plates a scatter of the maximal buckling load factor for cylinders is also observed – see Figure 4 but the location of the optimum (at the boundary) results in much better correlation with the case as the continuous variables (5) are used in the analysis. The plot has been obtained after 100 runs of the Evolver package using the classical GA as the optimization algorithm. The number of optimum results corresponding to the exact one is much higher than in the previous case of stacking sequence optimization for multilayered plates.

5. Discussion of the results

The possibility of a general approach to various types of discrete stacking sequence optimization problems with the use of discrete and continuous design variables have been formulated and studied. The presented results demonstrate the effectiveness of the proposed method of the solution with the use of the classical discrete design variables combined with the Evolver package. The accuracy of the results has been compared with those obtained by Haftka [10]. However, the above analysis shows evidently that the use of the classical GA is not always very convenient for our optimization problem (2)-(4) since it is impossible to answer whether the global optimum is reached or not. As it may be observed as the number of constraints increases the effectiveness of the method decreases – compare Figure 1 and Figure 4. Therefore, one can conclude that it is better to change the optimization algorithm but in our opinion it is much better to replace the ordinary set of discrete design variables by a continuous one introduced by equation (1). Briefly speaking it results in the replacement of the wheel roulette method by the elite one. In fact such a change of design variables, according to the nomenclature used in the optimization problems, is equivalent to the introduction of a new algorithm called as evolutionary algorithms.

6. Conclusion

The proposed method of the solution with the use of the classical discrete design variables combined with the Evolver package is an effective and it results in the replacement of the wheel roulette method by the elite one. A change of design variables, according to the nomenclature used in the optimization problems, is equivalent to the introduction of a new algorithm called as evolutionary algorithms, and the obtained results can be used to solution similar optimization problems.

7. References

[1] Haftka R T 1999 Genetic Algorithms for Optimization of Composite Laminates *Mechanics of Composite Materials and Structures* NATO Sc. Series, Vol. .361, Kluwer Academic Publ., pp. 430-442

[2] Hajela P, Gurdal Z and Haftka R T 1999 Design and Optimization of Laminated Composites *John Wiley and Sons* New York

[3] Muc A 2018 Evolutionary design of engineering constructions *Latin American Journal of Solids and Structures* 15 (4) art. no. e87 p 21

[4] Muc A 2016 Choice of Design Variables in the Stacking Sequence Optimization for Laminated Structures *Mechanics of Composite Materials* 52 (2) pp. 211-224

[5] Muc A and Ulatowska A 2012 Local fibre reinforcement of holes in composite multilayered plates *Composite Structures* 94 (4) pp. 1413-1419

[6] Song Z G, Zhang L W and Liew KM 2016 Optimal shape control of CNT reinforced functionally graded composite plates using piezoelectric patches *Composites B*. 85, 140–149

[7] Asadi H 2018 Numerical simulation of the fluid-solid interaction for CNT reinforced functionally graded cylindrical shells in thermal environments *Acta Astronaut* 138 pp. 214-224

[8] Miki M 1986 Optimum design of fibrous laminated plates subjected to axial compression *Proc. 3rd Japan-US composite Materials Conf.* Tokyo pp. 673-81

[9] Vinson J 1993 The behaviour of shells composed of isotropic and composite materials *Kluwer*

[10] Haftka R T and Le Riche R 1993 Optimization of laminate stacking sequence for buckling load maximization by genetic algorithm *AIAA Journal* 31 ss. 951-56

IOP Conf. Series: Materials Science and Engineering **893** (2020) 012008 doi:10.1088/1757-899X/893/1/012008

The Influence of the Pitch to Diameter Ratio (P/D) on a Novel Tidal Turbine Performance

Roslynna Rosli[1], Rosemary Norman[2], and Mehmet Atlar[3]

[1] Universiti Teknologi Brunei, Bandar Seri Begawan BE 1410, Brunei Darussalam
[2] Newcastle University, Newcastle upon Tyne NE1 7RU, United Kingdom
[3] University of Strathclyde, Glasgow G4 0LZ, United Kingdom

E-mail: roslynna.rosli@utb.edu.bn

Abstract. Computational Fluid Dynamics (CFD) offers numerical modeling investigating the performance that can be used other than or in tandem with experimental investigation. In order to offer meaningful results, the grid sensitivity test or Grid Convergence Index are usually carried out to ensure the solutions converging. This paper presents both grid sensitivity test results and the GCI calculation of a same numerical model exercised to investigate the performance of a tidal turbine, the Hydro-Spinna, with different P/D ratio. The CGI calculation presented the fine grid relative error to be 2.34% for the power coefficient and 2.28% for the thrust coefficient at their optimal TSR respectively. It was found that turbine with lowest P/D ratio has the highest power and thrust coefficient as well as TSR operational range. The turbine with P/D = 0.43 gives a power coefficient of 0.32 at the optimal TSR of 2.25.

1. Introduction

1.1. Introduction

The development in renewable energy specifically tidal energy is steadily progressing with aim to increase the contribution of renewable energy resources. Added with the critical issue of climate change, the research and development of tidal energy technologies are definitely welcome. The European Marine Energy Centre (EMEC) [1] in Scotland offers a hub for the development of tidal and wave energy technologies with different prototypes deployed in real sea conditions. In addition, the potential of new tidal energy technologies can also be investigated with a smaller scaled model in laboratory facilities such as cavitation tunnels and towing tanks. Investigation using Computational Fluid Dynamics can also be performed offering a vast platform for investigation especially during the early stage of design.

The Hydro-Spinna is a tidal turbine designed by Michael Gilbert and developed at Newcastle University. The Hydro-Spinna turbine extends in the axial direction as illustrated in Figure 1 making the turbine more three dimensional than a typical horizontal axis turbine. In order to assess the power potential of the turbine, the influence of the pitch length on the turbine performance needs to be assessed first. Wen [2] calculated that a lower P/D produced higher power using Blade Element Momentum Theory (BEMT) with the assumption that the turbine is a thin actuator disc. In order to investigate this, a numerical model was developed to predict the power and thrust coefficients of the turbine.

1.2. Hydro-Spinna Turbine

The Hydro-Spinna turbine is a novel horizontal axis tidal turbine presenting a helicoidal blade spiraling around the main axis as shown in Figure 1. The blades possess a cardioidal shape as illustrated in [3]. The HS500 is a 500 mm diameter model of the Hydro-Spinna turbines with cardioid shaped leading and trailing edge. The extended description and detailed geometry of the Hydro-Spinna HS500 model can be found in [3, 4]. The key parameters of the three turbines investigated are given below in Table 1.

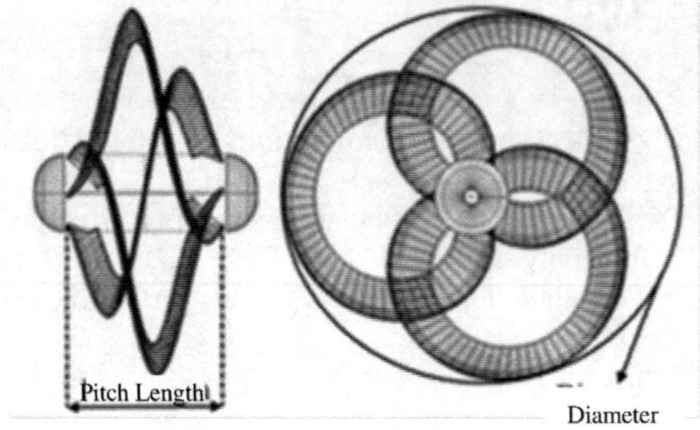

Figure 1. The Hydro-Spinna turbine parameters

Table 1. The turbine parameters with different P/D ratio

Pitch to Diameter Ratio, *P/D*	Diameter, D (mm)	Pitch, *P* (mm)	Pitch Angle at Root Section, *β* (Deg)	Hub Diameter, D_H (mm)
0.43		214	50.82	
0.75	500	375	60.30	100
1.00		500	68.11	

2. Numerical model

2.1. Reynolds-Averaged Navier Stokes Equations (RANS)

RANS and the continuity equations as defined in equations (1) and (2) were employed for the numerical model. RANS model is a modified Navier Stokes equation (3) taking the time averaged values of the fluctuating velocities and pressure variables to solve the equations. RANS was used due to its less computational time compared to Large Eddy Simulation models [5].

$$\frac{\partial(\rho\bar{V})}{\partial t} + \rho\left(\bar{V}.\vec{\nabla}\right)\bar{V} = -\vec{\nabla}.p + \mu\vec{\nabla}^2\bar{V} + \vec{\nabla}T_t + f_b \quad (1)$$

$$\frac{\partial\rho}{\partial t} + \nabla.(\rho V) = 0 \quad (2)$$

$$\frac{\partial(\rho V)}{\partial t} + \rho\left(V.\vec{\nabla}\right)V = -\vec{\nabla}.p + \mu\vec{\nabla}^2 V + f_b \quad (3)$$

In addition, the hybrid Menter SST k-ω turbulence model is employed to consider the advantage of the k-ω at the near wall and k-ε in the fully turbulent region [6]. The Menter SST k-ω model are defined in equation (4) and (5) respectively.

$$\frac{\partial}{\partial t}(\rho k) + \nabla.(\rho k\bar{V}) = \nabla.\left[\left(\mu + \frac{\mu_t}{\sigma_k}\right)\nabla k\right] + P_k + \beta^*\rho k\omega \quad (4)$$

$$\frac{\partial}{\partial t}(\rho\omega) + \nabla.(\rho\omega\bar{V}) = \nabla.\left[\left(\mu + \frac{\mu_t}{\sigma_{\omega,1}}\right)\nabla\omega\right] + P_\omega - \beta_2\rho\omega^2 + 2\frac{\rho}{\sigma_{\omega,2}\omega}\frac{\partial k}{\partial x_k}\frac{\partial\omega}{\partial x_k} \quad (5)$$

2.2. Numerical domain

The numerical domain independency test was done where the dimension of 3D-8D-3D were found to be the optimal dimension. This means that inlet and outlet are 3D and 8D away from the centre of the turbine respectively, whereas the radius of the cylindrical domain is at 3D from the centre of the turbine. The domain is divided into two parts i.e. the stationary and rotating domain where Moving Reference Frame was used for the rotating domain to represent the rotation of the turbine as shown in Figure 2. The rotating domain was set to be at 1.5D from the turbine centre, with the upstream and downstream boundaries set at 2D distance.

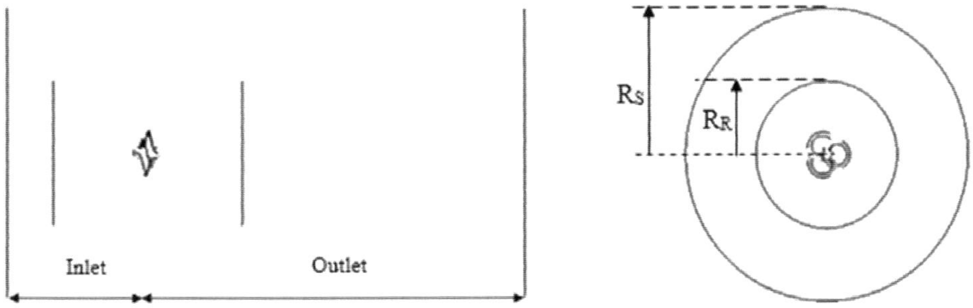

Figure 2. The numerical domain where inner vertical lines represent the upstream and downstream boundary of the rotating domain. R_R and R_S is the radius of the rotating and stationary domain respectively

2.3. Mesh independency test

The mesh independence analysis was conducted with the settings listed in Table 2 below. As mentioned previously, the domain dimensions for all model was set at 3D-8D-3D.

Table 2. Mesh settings of the numerical domain

Mesh Setting	Rotating (Rotating)	Stationary (Stationary)	Total
		Mesh Number	
1	1,439,663	216,322	1,655,985
2	1,847,224	216,322	2,063,546
3	1,847,224	355,116	2,202,340

2.4. Grid Convergence Index (GCI)

The numerical model was also validated using GCI where the procedure used in this paper is based on the widely accepted Richardson Extrapolation (RE) [7 -9] obtained from obtained from Celik et al [10] as described below.

Firstly, a representative mesh cell size, h, for three dimensional calculations is calculated as defined in equation (6).

$$h = \left[\frac{1}{N}\sum_{i=1}^{N}(\Delta V_i)\right]^{1/3} \qquad (6)$$

where ΔV_i is the volume of the ith cell, and N is the total number of cells used in the computations.

Then, select three different sets of grids and run simulations to determine the value of the solutions for the study i.e. in this case power and thrust coefficients which is denoted as variable ϕ. The refinement factor $r = h_{coarse}/h_{fine}$ is recommended to be greater than 1.3. Let $h_1 < h_2 < h_3$ and $r_{21} = h_2/h_1$, $r_{32} = h_3/h_2$ and calculate the apparent order p of the method using equation (7) to (9).

$$p = \frac{1}{ln(r_{21})}|ln|\varepsilon_{32}/\varepsilon_{21}| + q(p)| \qquad (7)$$

$$q(p) = ln\left(\frac{r_{21}^{p}-s}{r_{32}^{p}-s}\right) \qquad (8)$$

$$s = 1.sgn(\varepsilon_{32}/\varepsilon_{21}) \qquad (9)$$

where $\varepsilon_{32} = \emptyset_3 - \emptyset_2$, $\varepsilon_{21} = \emptyset_2 - \emptyset_1$ and $\emptyset_k$ represents the solution on the kth grid.

Calculate the grid convergence index as defined in equation (12) by first calculating the approximate relative error defined in equation (10) and the extrapolated relative error as defined in equation (11).

$$e_a^{21} = \left|\frac{\emptyset_1 - \emptyset_2}{\emptyset_1}\right| \tag{10}$$

$$e_{ext}^{21} = \left|\frac{\emptyset_{ext}^{12} - \emptyset_1}{\emptyset_{ext}^{12}}\right| \tag{11}$$

$$\mathrm{GCI}_{fine}^{21} = \frac{1.25 e_a^{21}}{r_{21}^{p} - 1} \tag{12}$$

3. Results and discussion

3.1. Validation of numerical model

The mesh sensitivity test considered three different sizes as indicated in Table 2. The results among the three settings were consistent with no significant difference as shown in Figure 3. The consistency in the mesh results was probably due to the small difference in the number of mesh with only a difference of approximately 200,000 cells between each setting. The results from the independency test presents a reliable numerical model for the turbine performance investigation. Similarly, other researchers also obtained optimum mesh numbers of approximately 2 million cells [11, 12]. In addition, the CGI calculation as shown in Table 3 resulted the numerical uncertainty for the fine-grid solution was 2.34% for the power coefficient and 2.28% for the thrust coefficient. Now that the numerical model has been validated both by using mesh sensitivity test and the GCI calculation, it is further used to investigated the performance of the turbine at different P/D ratio.

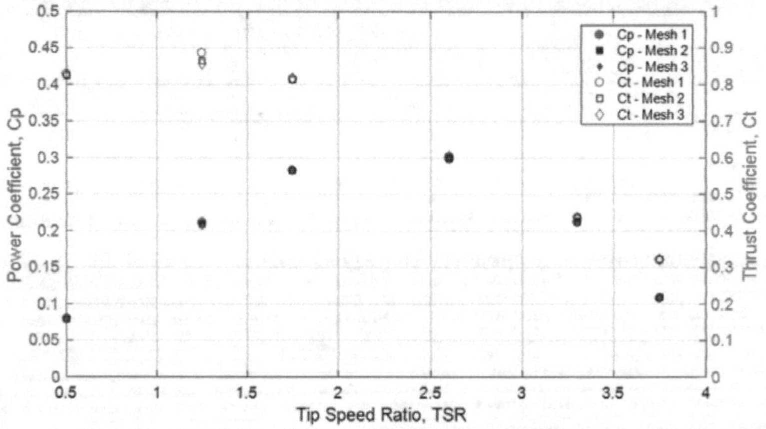

Figure 3. Mesh sensitivity result of the numerical model

Table 3. GCI results for the numerical model used

	ϕ = power coefficient at TSR = 2.6	**ϕ = thrust coefficient at TSR = 1.75**
N_1, N_2, N_3	2202340, 2063546, 1655985	
r_{21}	1.021	1.021
r_{32}	1.076	1.076
ϕ_1	0.2982	0.8144
ϕ_2	0.2991	0.8165
ϕ_3	0.2978	0.8136
p	7.173	6.355

e_a^{21}	0.0030	0.0026
e_a^{32}	0.0043	0.0036
$\mathbf{GCI}_{coarse}^{32}$	0.79%	0.75%
$\mathbf{GCI}_{fine}^{21}$	2.34%	2.28%

3.2. Turbine performance

The influence of the P/D ratio on the Power Coefficient (C_P) and Thrust Coefficient (C_T) was investigated for a full range of operational Tip Speed Ratio (TSR). The equations are defined below for reference.

$$C_P = \frac{Q\Omega}{0.5\,\rho A U^3} \tag{13}$$

$$C_T = \frac{T}{0.5\,\rho A U^2} \tag{14}$$

$$TSR = \frac{\Omega R}{U} \tag{15}$$

where Q is the torque, T is the thrust, Ω is the angular velocity of the turbine, ρ is the density of the fluid, A is the swept area of the turbine, and U is the flow velocity.

It was found that the P/D ratio greatly varies the performance characteristic of the turbine where the higher the ratio, the lower the power coefficient as shown in Figure 4. The results were found to be in agreement to the work conducted by Wen [2] where lower P/D gives the highest performance. The Hydro-Spinna turbine with P/D = 0.43 gives a power coefficient of 0.32 at the optimal TSR of 2.25. The operational range of this turbine is $0 < TSR < 4.15$. The higher the pitch length is the further the turbine is extended in the axial direction, while the diameter stays constant. Therefore, the higher the P/D ratio, the pitch angle of the turbine blade also increases hence contributing to the low power coefficient.

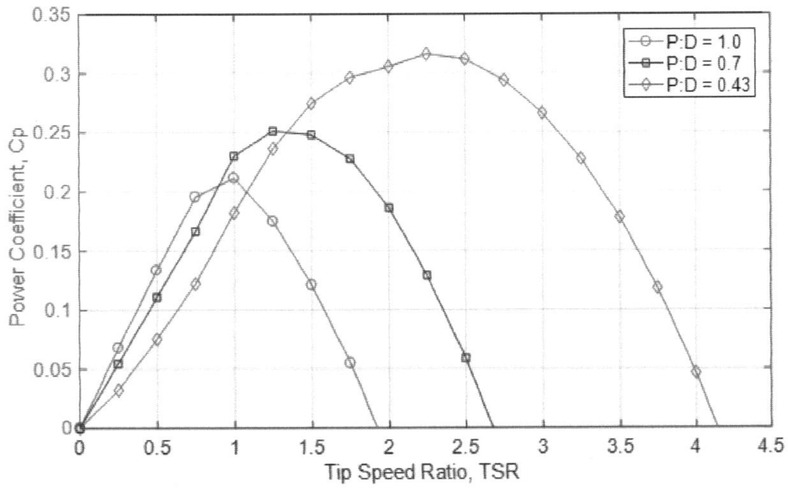

Figure 4. The power coefficient of the Hydro-Spinna turbine with different P/D ratio

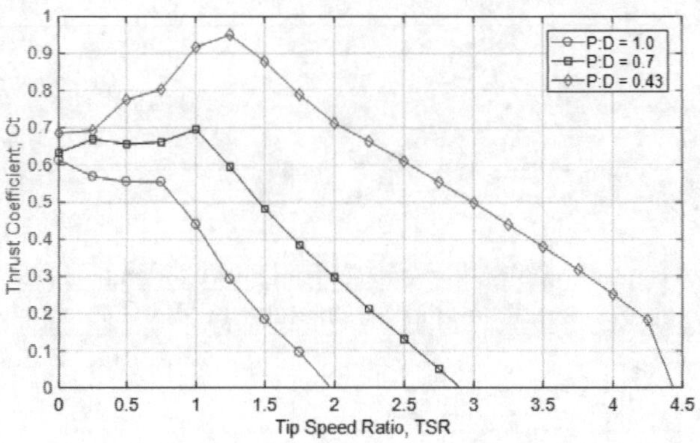

Figure 5. The thrust coefficient of the Hydro-Spinna turbine with different P/D ratio

In addition, it was also concluded that while the power increases with lower P/D the operational TSR range also increases. Where turbine with P/D ratio of 1 has a range of $0 < TSR < 1.9$. Hence, the Hydro-Spinna turbine with the lowest P/D ratio gives the highest power coefficient. Consequently, it also has the highest thrust coefficient among the three as presented in Figure 5. The turbine with P/D of 0.43 has a maximum thrust coefficient of 0.95 at TSR of 1.25, whilst at its optimal TSR i.e. TSR = 2.25, the thrust coefficient was only 0.66, 30% less than the maximum thrust. The high thrust at TSR = 0 is believed to be caused by the high solidity of the turbine surface blocking the current flow.

4. Conclusion

The performance characteristic of the Hydro-Spinna with different P/D ratio was successfully investigated by using a numerical model. The numerical model was validated by conducting domain and mesh sensitivity test as well as by GCI calculation. The numerical model validations were conducted to ensure the credibility of the model used. The numerical investigations concluded that that turbine with lowest P/D ratio has the highest power and thrust coefficient. Similarly, the operational *TSR* range of the turbines decreases with higher ratio.

5. References

[1] EMEC 2013 European Marine Energy Centre, Orkney, Scotland
[2] Wen Y. 2011 Optimisation and experimental validation of a novel marine turbine device. Newcastle University thesis.
[3] Rosli R, Norman R, and Atlar M 2014 GRAND Renewable, Tokyo, Japan, 30th July – 1st August 2014.
[4] Rosli R, Norman R, and Atlar M 2016 *Renewable Energy* **99** 1227-34.
[5] Breuer M, Jovičić N, and Mazaev K 2003 *International Journal for Numerical Methods in Fluids* **41(4)** 357-88.
[6] Menter FR 1996 *Journal of Fluids Engineering* **118(3)** 514-9.
[7] Roache PJ 2003 *Journal of Fluids Engineering* **125(4)** 731-2.
[8] NPARC Alliance 2008 NPARC Alliance CFD Verifications and Validation Website
[9] Ali MSM, Doolan CJ, and Wheatley V 2009 7th International Conference on CFD in the Minerals and Process Industries, Melbourne, Australia, 9th - 11th December 2009.
[10] Celik IB, Ghia U, Roache PJ, Freitas CJ, Coleman H, and Raad PE 2008 *Journal of Fluids Engineering* **130(7)** 0780011-4.
[11] Bai CJ, Hsiao FB, Li MH, Huang GY, and Chen YJ 2013 *Procedia Engineering* **67** 279-87.
[12] Edmunds M, Williams AJ, Masters I, and Croft TN. 2017 *Renewable Energy* **101** 67-81.

Acknowledgments
The authors would like to acknowledge the creative mind of Mr Michael Gilbert, the Hydro-Spinna creator.

IOP Conf. Series: Materials Science and Engineering **893** (2020) 012009 doi:10.1088/1757-899X/893/1/012009

Storage Allocation For Stocked and Buffer Baskets

Arunrat Walitsarangkul and Oran Kittithreerapronchai

Department of Industrial Engineering, Chulalongkorn University, 10330 Bangkok, Thailand

Abstract. The recent changes in the retailer landscape and the continuous labor shortage in Thailand and have caused many distribution centers to embrace material handling equipment capable of automatically storage and transporting products. This article presents the case study of a home improvement retailer that adopts a vertical lift module (VLM) system into one of its distribution centers. As automatic equipment dedicated for baskets, a VLM system transports two types of baskets, particularly available stock baskets preparing for picking and consolidated baskets waiting for shipping. A preliminary analysis of current operations reveals that the number of baskets allocated for each type is not suitable, resulting in low utilization of the equipment and storage space. After studying the historical data as well as nature of the business, we proposed a simulation to experiment with the effects of the following operating policies (1) the suitable number of baskets allocated for each type (2) storage policy of consolidated baskets and (3) retrieving patterns of consolidated baskets. The results suggest that the suitable level of each factor that reduces the space utilization of 6.67% and vehicles travel time of 37.5%. Nevertheless, travel times of incoming and outgoing lifters are close to similar represented by their utilization.

1. Introduction

For a few decades, the expansion of domestic infrastructure projects and the growth of their adjacent construction projects have fueled the construction material and home improvement retailers in Thailand. Nevertheless, many retailers have contemplated their distribution strategies as the recent changes in consumers' behavior and technological developments have shifted the landscape [1,2]. Coupled with an aging society, they have embraced material handling equipment capable of automatically storage and transporting products as a means to overcome labor shortage and high wages [3,4].

The article presents the case study of a home improvement retailer that has faced with the challenges mentioned above and recently reconfigured its distribution center with a semi-automation system. This state-of-art distribution center is and equipped with different material handling, such as automatic storage and retrieval system (AS/RS), conveyor systems, as well as vertical lift module (VLM) system. The VLM consists of lifters and vehicles transporting baskets between the conveyor system and basket-size selective racks. As a dedicated system, the baskets can be classified into two types, particularly available stock baskets preparing for picking and consolidated baskets waiting for shipping.

As the first site in Thailand that adopted such equipment, the warehouse manager concerns as the previous researches suggested that the productivity of the system depends on operating policies. Furthermore, a preliminary analysis of current operations reveals that the number of baskets allocated for each type is not suitable, resulting in low utilization of the equipment and storage space. As a result, we studied the historical data and nature of business and embedded it into a discrete-event simulation.

The proposed simulation was designed to experiment with the effects of the following operating policies (1) the suitable number of baskets allocated for each type (2) storage policy of consolidated baskets and (3) retrieving patterns of consolidated baskets. After the relevant literature review in Section 2, the remaining sections of this article are organized as follows. Section 3 highlights the background of the case study distribution center and the main configuration of VLM and its related systems. Having verified input data and operational information, Section 4 focuses on the components and interactions of the VLM system as well as its represented simulation model. Before suggesting suitable operational policies in Section 6, the results of the simulation model that experiments with different scenarios are analyzed and discussed in Section 5.

2. Literature review

Related literature can be grouped into two categories: Vertical Lift Module and Warehousing policies.

2.1. Vertical lift module

As dedicated equipment designed for uniform and standardized objects, the VLM system integrates automatic transporting of objects with manual operations. Because of its potential labor reduction and independent transporting of lifters and vehicles, many researchers have elaborated and compared the VLM system with other material handling systems. Toor [5] proposed a VLM system instead of a manual picking of fast-moving products as it reduced the time required to complete an order. Meller [6] created a model to compare a VLM system with a carousel for batch picking and found that a VLM provided higher throughput, especially with fast-moving products. Later, Lenoble [7] applied a mathematical model to compare picking policies and to determine a suitable grouping that minimizes the total picking time. The study reveals that grouping orders in the same tray result in the lowest picking time. In terms of economics, Sgarbossa [8] compared a VLM system and a manual carton system and found that the installation could reduce the total annual cost. Calzavara [9] extended the study by using an industrial case study and found that using the VLM system could reduce worker costs.

2.2. Warehousing policies

As the most labor-intensive activity [10], a good picking process requires the developing of suitable storage rules and picking policies. In general, there are three storage rules, namely random storage, dedicated storage, and class-based storage. Despite the space utilization benefit of random storage, Choe and Shape [11] concluded that the selection of suitable storage rules involves many other factors. In terms of picking policies, a typical warehouse usually applies a combination of single order picking, zone picking, and batch picking to different groups based on physical products, demand patterns, size of storages, and available manpower [12, 13].

3. Home improvement distribution center

As an essential part of the home improvement retailer, the case study distribution center provides dedicated warehousing and distributing services to its distribution network. Located in the north of Bangkok, the distribution center consists of many warehouses categorized by merchandized classes within the same area. One of which is the newly constructed warehouse using the 'goods-to-men' concept that stores consumer-good and home appliance categories. In addition to high-value items, this warehouse also serves as a merge-in-transit facility that combines cross-docking items received from domestic supplies with storage items purchased from international vendors. Because of the values of shipments, the semi-automated warehouse emerges as the center of warehousing operations.

3.1. Basket operation

Among different units of handling, the basket operation is viewed as one of the most critical and the focal points of the design as it connects different systems and integrates cross-docking item with storage ones. The basket system consists of three components, as follows:

- **Basket selective rack** is the storage equipment in the system that consists of two aisles and four storage banks. The rack has 18 levels and 22 bays with a storage capacity of 7,920 baskets. The initial configuration divides the storage capacity into 2,500 buffer baskets and 5,420 stock baskets.
- **Lifter** is the material handling equipment that vertically connects the basket conveyor system with vehicles. It can transfer two baskets per trip and travels at speed up to 100 meters per minute. The equipment is further classified by its function into incoming lifters and outgoing lifters. Each lifter operates independently from one other in its aisle.
- **Vehicle** is the material handling equipment that horizontally transfers baskets between lifters and baskets selective rack at speed up to 200 meters per minute. Each level has its own vehicle or 36 units in total.

3.2. Operating time of lifter and vehicles in VLM system

Because of its automatic transporting, the operating time of a lifter and a vehicle is consistent and consists of three components: communicating time, handling time, and traveling time. Based on operation logs, the communicating times of a lifter and a vehicle are constant at 3.01 and 3.91 seconds, respectively. Because a lifter may carry up to two baskets in the same trip, its handling times depend on a number of baskets, particularly 1.4 and 2.5 seconds for the first basket and the second basket, respectively. Conversely, the handling time of the vehicle is a constant at 8.2 seconds. In general, the traveling times directly depends on the position of an origin and a destination. Specifically, the traveling time of a lifter and a vehicle is a linear function and a piecewise linear function, as shown in Equations 1 and 2:

$$T^{\text{lifter}}(d) = 0.6497 + 0.2807 \cdot d, \qquad d \neq 0 \tag{1}$$

$$T^{\text{vehicle}}(d) = \begin{cases} 2.09 \cdot d & \text{, if } d \leq 1 \\ 2.2947 + 0.2779 \cdot d & \text{, if } 2 \leq d \leq 10 \\ 3.3552 + 0.1553 \cdot d & \text{, if } d > 10 \end{cases} \tag{2}$$

where d is a position of an origin and a destination and T^{lifter} and T^{vehicle} are traveling times of a lifter and a vehicle, respectively.

3.3. Nature of demand

Because stocked items consist of more than 3,000 articles, each of which has a different physical dimension and demand pattern. These factors affect the numbers of an item available in each basket as well as picking frequency. Therefore, we applied Pareto analysis and classified articles based on numbers available in a basket and average demand, as shown in Figure 1.

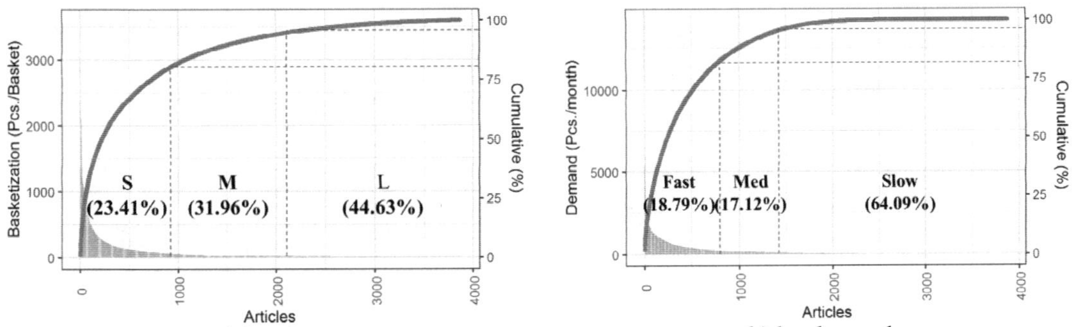

a) by size b) by demand

Figure 1. Pareto analysis of articles by their physical sizes and requested demands

IOP Conf. Series: Materials Science and Engineering **893** (2020) 012009 doi:10.1088/1757-899X/893/1/012009

These classifications yield the division of total articles into nine categories as shown in Table 1. Before embedded in a simulation model, the generated distribution of the categories are, as shown in Figure 2.

Table 1. Number of articles of each category

Piece/basket		Demand		
		Fast (≥ 270)	**Med** (90-270)	**Slow** (≤ 90)
S	(≤70)	234	132	541
M	(18 - 70)	225	230	783
L	(≥ 70)	269	301	1,159

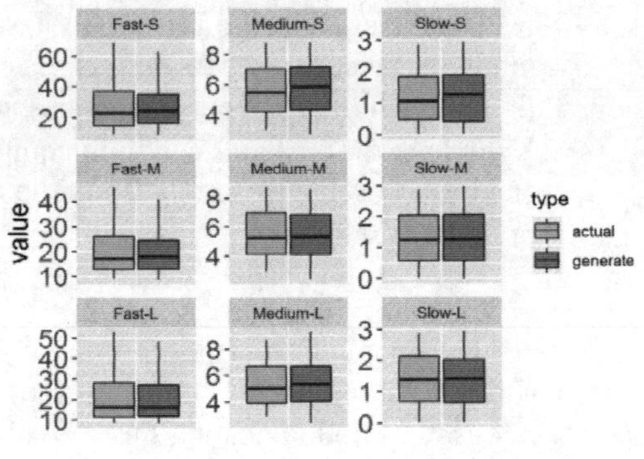

Figure 2. Comparsion the fits of distributions with actual data

The validation of generated categories and actual ones with two-sample Kolmogorov-Smirnov test reveals no statistical difference at the 0.95 confidence level.

4. Simulation model

Before explaining a simulation model, it is important to understand the operation, as shown in Figure 3.

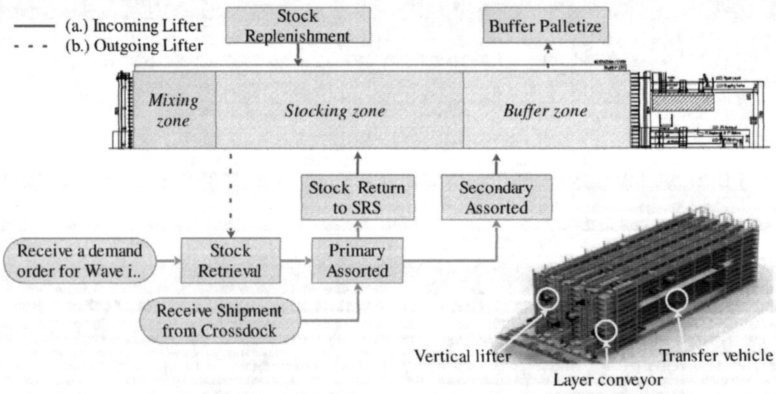

Figure 3. Operation process flow

4.1. Operations process in the VLM system

Stocked items are replenished into the VLM system by cartons from another storage area and arranged into baskets before stored at the stocking zone of the basket selective racks. At the beginning of each picking wave, specific baskets of requested items are retrieved and transferred to primary sorted stations. At the stations, items are assorted into several transport baskets destined for eight different secondary assorted stations based on requested quantities. At the destination, transport baskets are merged with cross-docking baskets and picked into different buffer baskets presenting the same stores by a worker. Once accumulated items in a buffer basket reach a certain height, the conveyor belt transfers the basket into the buffer zone of the selective rack waiting for the accumulation of 25 baskets or the end of picking wave. These outgoing baskets are sequentially requested by a store at buffer palletizing stations before stacking on a pallet and transporting to its own docking area.

4.2. Simulation model
As the most suitable tool to model and analyze a complex system, a discrete-event time simulation is selected to provide insights and compare the combination of policies without disrupting the ongoing operation. The simulation is organized similar to the current operation discussed in Section 4.1 that schedules into three six-hour slots.

4.3. Operational policy
With the exception of urgent picking and replenishment, the retailer currently considers a scenario that consists of the combination of storage policy and retrieval policy as follows:

Storage policy represents the selection of storage locations in a selective rack by a different type of baskets. This policy can be further divided into two alternatives, particularly

- **Two storage zones** (S0): This is the initial configuration that reserves continuous buffer baskets near lifters, called buffer zone, at the maximum requirement or 25 baskets per store, thereby ensuring no blocking. The remaining and further storage capacity is referred to as the stocking zone and allocated to stock baskets.
- **Three storage zones** (S1): Since some retail branches are more active in a particular wave, the alternative proposed mixing zone where stock baskets and buffer basket can occupy if their zones are full. Based on the physical configuration of rack, buffer, stocking, and mixing zones, sequentially consist of 20, 80, and 10 bays, respectively.

Retrieval policy reflects a logical setting in which stock baskets of a particular product are retrieval. The policy can be further categorized into three alternatives.

- **First-In-First-Out** (R0): As the initial configuration, the alternative sequences stock baskets based on the time stamp in the system.
- **Pick-to-Clear** (R1): The alternative selects the stock basket with the smallest quantities first as it aimed to generate more storage locations.
- **Pick-at-Most** (R2): A popular alternative that sequences stock baskets based on their quantities and selects a basket whose product quantity is more than and/or close to requested quantities first.

5. Results and discussion
To compare these scenarios, we embedded related parameters and operations into a simulation using R/RStudio [14]. Each combination and repeated 30 replications to obtain 10% relative precision [15] with seven days time horizon. The key results of the simulation include space utilization, lifter travel time, and vehicle travel time, as shown in Table 2.

Table 2. Space utilization, lifter travel time and vehicle travel time from the simulation model

Scenario	Policy		Space Utilization	Average Travel time (minute/day)		
	Space Allocation	Retrieval Sequencing		Incoming Lifter	Outgoing Lifter	Vehicle
0	S0	R0	41.99%	410.17 ± 38.57	432.34 ± 53.66	1,090.26 ± 89.19
1	S0	R1	39.20%	410.30 ± 38.44	433.90 ± 52.94	1,090.34 ± 99.45
2	S0	R2	42.44%	397.71 ± 38.28	422.48 ± 52.81	1,042.70 ± 100.68
3	S1	R0	42.00%	416.07 ± 39.21	444.06 ± 54.23	804.16 ± 58.32
4	S1	R1	39.20%	416.12 ± 39.12	445.34 ± 53.72	792.76 ± 63.04
5	S1	R2	42.41%	403.69 ± 38.99	434.48 ± 53.76	755.05 ± 60.71

In Table 2, Scenario 1-5 are used to probe different policies, whereas Scenario 0 serves as a base case to compare. In terms of space allocation policy, the result reveals that three-storage-zones alternative dominates the two-storage-zones alternative as baskets are stored closer to the P/U station, resulting in a reduction of vehicles travel time. The effects of the retrieval sequencing policy can be found when comparing Scenario 0 to Scenario 2 as the pick-to-clear alternative providers a lower space utilization lower than the FIFO alternative and the pick-at-most alternative. Nevertheless, travel times of incoming and outgoing lifters are more or less similar represented by their utilization, as shown in Figure 4.

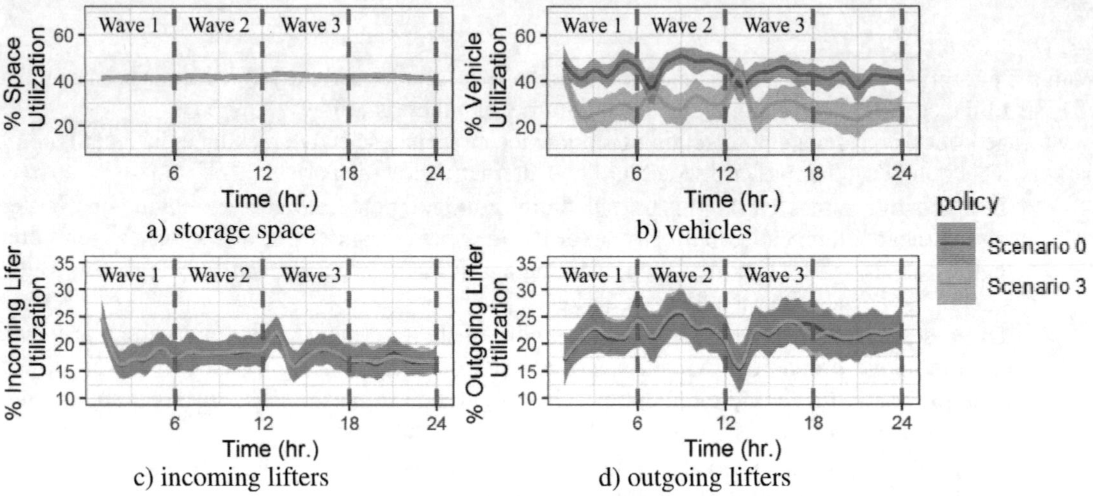

Figure 4. Utilizations of storage and material handling equipment in Scenario 0

The average daily utilization and its variance in the VLM system are represented as solid lines and shaded areas, respectively. Except for the end of wave (dashed lines), the daily space utilization is stable as numbers of incoming baskets and outgoing baskets are approximately equal. At the end of waves, the space utilization is fluctuated due to simultaneously requesting buffer baskets. This reason aligns with the utilization of vehicles in Figures 4.b that the utilization of vehicles increases because of replenishment as well as relative distance of stocked baskets. Because of fewer stores in the third wave, the utilization of vehicles is relatively unchanged. Figures 4.c and 4.d represent the utilization of incoming lifters and outgoing lifters, which exhibit the opposite pattern. Particularly, the high (low) utilization of incoming (outgoing) lifters is caused by stock baskets. Conversely, the utilization of incoming (outgoing) lifters decreases (increases) at the end of waves as buffer baskets are requested.

6. Conclusion

This study reported an ongoing effort to adopt automated material handling equipment in Thailand and the adaptation of the VLM system in a distribution center of home improvement retailers. After analyzing and simplifying the related products and processes, we embedded empirical parameters into a simulation model and experimented with retrieving policy and storage policy. The combinations of these policies can be grouped into six different scenarios in which the percentage of space utilization, lifter traveling time, and vehicle travel time are computed and compared. The result reveals that the most suitable combination of operational policies is the scenario in which selects three zones and pick-to-clear as storage and retrieval policies, respectively. With a moderate utilization of the system, the scenario provides relatively low traveling time of vehicles while minimizes the number of baskets requested. However, the further system may require if the retailer decides to expand its branches or engage in the online market. The suggestion of utilizing three zones storage policy also raises questions on the sizes of each storage zone and the possibility of overlapped zones.

7. References

[1] Klinchuanchun, P. (2018). *Thailand Industry Outlook, Housing in Bmr.* Retrieved from Krungsri Research: https://www.krungsri.com/bank/en/Other/research/industry/industry-outlook.html

[2] Toomwongsa, N. (2018). *Thailand Industry Outlook, Construction Contractor*. Retrieved from Krungsri Research: https://www.krungsri.com/bank/en/Other/research/industry/industry-outlook.html
[3] Leepipatpaiboon, P. & Thongsri, N. (2017). *Industrial Robots and Its Impacts on Labor Market*. Retrieved from Bank of Thailand: https://www.bot.or.th/Thai/MonetaryPolicy/Pages/default.aspx
[4] Office of the National Economic and Social Development Council. (2013). *Population Projections for Thailand 2010-2040*. Retrieved from Social and Quality Of Life Database System: http://social.nesdb.go.th/social/
[5] Toor, P. S. (2015). *Improving operational efficiency of a semiconductor equipment manufacturing warehouse through strategic allocation of parts*. (Doctoral dissertation), Massachusetts Institute of Technology.
[6] Meller, R. D., & Klote, J. F. (2004). A throughput model for carousel/VLM pods. *IIE Transactions, **36**(**8**)*, 725-741.
[7] Lenoble, N., Frein, Y., & Hammami, R. (2016). *Optimization of order batching in a picking system with a Vertical Lift Module*. Paper presented at the International Conference on Information Systems, Logistics and Supply Chain.
[8] Sgarbossa, F., Calzavara, M., & Persona, A. (2018). Economic and Performance Analysis of Dual-bay Vertical Lift Modules. [9] Calzavara, M., Sgarbossa, F., & Persona, A. . (2019). Vertical Lift Modules (VLMs) for small items order picking: an economic evaluation. *International Journal of Production Economics, **210***, 199-210.
[9] Calzavara, M., Sgarbossa, F., & Persona, A. . (2019). Vertical Lift Modules (VLMs) for small items order picking: an economic evaluation. *International Journal of Production Economics, **210***, 199-210.
[10] Hackman, S. T., Frazelle, E. H., Griffin, P. M., Griffin, S. O., & Vlasta, D. A. (2001). Benchmarking warehousing and distribution operations: an input-output approach. *Journal of Productivity Analysis, **16**(**1**)*, 79-100.
[11] Il-Choe, K., & Sharp, G. (1991). Small parts order picking: design and operation. https://www2.isye.gatech.edu/~mgoetsch/cali/Logistics%20Tutorial/order/article.htm
[12] Gu, J., Goetschalckx, M., & McGinnis, L. F. (2007). Research on warehouse operation: A comprehensive review. *European journal of operational research, **177**(**1**)*, 1-21.
[13] Gu, J., Goetschalckx, M., & McGinnis, L. F. (2010). Research on warehouse design and performance evaluation: A comprehensive review. *European journal of operational research, **203**(**3**)*, 539-549.
[14] Horton, N. J., & Kleinman, K. (2015). *Using R and RStudio for data management, statistical analysis, and graphics*: Chapman and Hall/CRC.
[15] Steiger, N. M., Lada, E. K., Wilson, J. R., Joines, J. A., Alexopoulos, C., & Goldsman, D. . (2005). ASAP3: A batch means procedure for steady-state simulation analysis. *ACM Transactions on Modeling and Computer Simulation (TOMACS), **15**(**1**)*, 39-73.

Acknowledgments
The authors would like to thank Mr.Chaiyuth Karunyasopon and Ms.Petchompoo Theppipidh for their time and valuable suggestions.

AUTHOR INDEX